Alfred T. Tucker Wise

Alpine winter in its medical aspects

with notes on Davos Platz, Wiesen, St. Moritz and the Maloja

Alfred T. Tucker Wise

Alpine winter in its medical aspects
with notes on Davos Platz, Wiesen, St. Moritz and the Maloja

ISBN/EAN: 9783742885999

Manufactured in Europe, USA, Canada, Australia, Japa

Cover: Foto ©berggeist007 / pixelio.de

Manufactured and distributed by brebook publishing software
(www.brebook.com)

Alfred T. Tucker Wise

Alpine winter in its medical aspects

MALOJA.

ALPINE WINTER

IN ITS MEDICAL ASPECTS:

WITH NOTES ON

DAVOS PLATZ, WIESEN, ST. MORITZ,

AND

THE MALOJA.

BY

A. TUCKER WISE, M.D., L.R.C.P., M.R.C.S,

*Formerly Visiting Physician to the Infirmary for Consumption, Margaret St., Cavendish Sq.;
Physician to the Western General Dispensary; Honorary Medical Officer to the
Kilburn and Maida Vale Dispensary;
House Physician, House Surgeon, and Resident Obstetric Officer, St. Mary's Hospital, London;
Member of the Harveian Society; Fellow of the Royal Meteorological Society, &c.;
Author of " Davos Platz, and the Effects of High Altitude on Phthisis;" " Wiesen as a Health
Resort in Early Phthisis;" &c., &c.*

FOURTH EDITION.

LONDON:

J. & A. CHURCHILL, NEW BURLINGTON STREET, W.

1884.

PREFACE.

THE present edition contains all the subject matter of previous publications of the Alpine Climate Series, with extracts from my papers read at the Harveian Society of London; the Royal Meteorological Society; and the International Medical Congress, held at Washington in September, 1887. A repetition of meteorological observations has been omitted. My endeavour in the following pages has been to represent Alpine climate in its true light (as experienced from personal observations extending over several years' residence), avoiding exaggeration or an omission of those details termed "drawbacks," so necessary to be portrayed in the consideration of any health-resort.

The treatment of chest diseases in the mountains rests on a solid basis of ascertained facts, there is consequently no need to strain matters nor hide all the blemishes of Alpine climate; it is but right that they should be expressed uncoloured and free from bias by those who know, for in this way only can any true appreciation of the climatic value of winter residence in these regions, be gained by medical men and others, who have neither the inclination nor the time for prolonged or complete observations.

A. T. W.

33 CURZON STREET, MAYFAIR,

September, 1888.

CONTENTS.

ILLUSTRATIONS.

INTRODUCTION.

THE first impression of a snow-clad landscape in a cold climate during fine weather is one of surprise and admiration. Conscious of excessive cold by external appearances rather than by disagreeable sensations of chilliness, one almost doubts the reality of the low temperature. A bright sun and blue sky overhead, a clear and quiet atmosphere, distant sounds transmitted to the ear through the still air, combine with the charms of the scenery to produce such buoyancy of spirits that a man is braced and invigorated for almost any exertion.

An Englishman inured to a damp and windy climate, and with an inherent love for outdoor sports and exercises, feels new life and energy as he views the glassy expanse of a frozen Canadian lake surrounded by dense forest, or makes his way through the rugged boulders and pines of Nova Scotia. A blazing sun above would almost persuade him it was summer, whilst he recognises the sing-song of the sleigh-bells in the distance, miles away it may be, and tastes the keen atmosphere of a bright winter's day.

The effects on the Anglo-Saxon race of living in a cold climate may be seen in Canada and some of the Northern States of America, where the race is physically superior and more vigorous than that of the parent stock. This, no doubt,

B

depends to a large extent on the mode of life, quality of food, and outdoor exercise taken, in addition to an artificial selection of species in emigration. Nevertheless, such is the fact that these cold climates, with wide ranges of temperature, harden the European constitution by invigorating the whole frame, and rendering the body less susceptible to changes in temperature, privation, or disease.

What the eventual results will be on the English race by the growth of population and consequent increase of indoor employment, it is difficult to foretell. It is pretty certain that dwellers in large towns, *employées* in factories, workshops, warehouses, offices, &c., are not exposed to the most favourable conditions of life in regard to health and robust development ; and although the "survival of the fittest" will select the strongest and most suitable being to withstand what may almost be termed the ravages of civilisation, even the selected one will scarcely be improved by the deteriorating effects ot over-crowding, impure air, improper food, want of sunlight and equable exercise of all the organs of the body.

With examples before us of the health-giving properties of cold regions, it may seem an oversight that cold climates as "change" never received the attention they merited until within recent years.

One of the first to foreshadow the track which is now being pursued by some of the leaders of medicine in England, France, and Germany, was Dr. Bodington, a practitioner in Warwickshire, who recommended "dry, frosty air" in the treatment of pulmonary consumption. In a small *brochure*, published by him forty-five years ago on this subject, he said : "To live in and breathe freely the open air, without being deterred by the wind or weather, is one important and essential remedy in arresting its progress—one about which there appears to have generally prevailed a groundless alarm less the consumptive patient should take cold . . . The abode of the patient should be in an airy house in the country—if on an eminence the better. The neighbourhood chosen should be dry and high ; the soil generally of a light loam, a sandy, or gravelly bottom. The atmosphere is in such situations comparatively free from fogs and dampness. The patient ought never to be deterred by the state of the weather from exercise in the open air ; if wet and rainy, a covered vehicle should be employed, with open windows. The cold is never too severe for the consumptive patient in this climate. The cooler the air which passes into the lungs · the greater will be the benefit the patient will derive. *Sharp, frosty days in the winter season* are most favourable. The application of cold pure air to the interior surface of the lungs is the most powerful sedative that can be applied, and does more to promote the healing and closing of cavities and ulcers of the lungs than any other means that can be employed . . . Many persons are alarmed and deterred from taking much exercise in the open air from the circumstance of their coughing much on their first emerging from the warm room of a house, but this shows that the air of the room was too warm, not that the common atmosphere was too cold." As adjuncts, he advocated the use of milk, wine, and generous diet. (*a*) In these remarkable sentences are clearly signalised the indications for residence at such health resorts as the high Swiss Valleys, and they anticipate in an unmistakable manner the treatment of pulmonary disease or weakness, by what is now termed somewhat euphoniously, " the high altitude cure."

This English practitioner appears to have been the first to introduce to notice as a therapeutic agent " dry, frosty, air," which has proved itself to be a remedy of practical value and application in the treatment of disease.

In more recent years Dr. Hermann Weber, with eventual

(*a*) " Essay on the **Treatment and Cure of Pulmonary Consumption**." By George Bodington. 1840.

success, drew attention to high, cold altitudes as presenting
favourable conditions during winter months for the arrest or
amelioration of phthisis. To him undoubtedly belongs the
credit of having brought the subject prominently before the
profession in a clear and intelligible manner, with results on
patients who had wintered in the mountains. In 1864 Dr.
Hermann Weber wrote : "The question of the influence of
Alpine Climate—*i.e.*, localities in Switzerland elevated beyond
5,000 feet above the level of the sea—on the tendency to
consumption, and on the very first stage of this disease, is one
of such vast importance that I cannot help asking for the
co-operation of the whole medical profession in its further
examination and decision." It may appear surprising that so
many years elapsed before the medical profession began to
take notice of climates with low temperature. In England
this can be readily accounted for, as probably the majority of
persons would judge a low temperature from their own experience of cold in the mother country, and winter time is
eminently unsuitable for medical men engaged in practice to
make any personal investigations into this or that health-
resort, therefore the hesitation evinced by the medical
faculty in the early years of the discussion on wintering
in the Alps, needs no comment or apology. For
many years Davos Platz has attracted large numbers
of *poitrinaires*, and from reports published by English
physicians at home the efficacy of mountain air in some
cases of phthisis is undeniable. The two pioneers who may
be said to have made the crucial experiment of testing the
effects of the icy winter air on the lungs were Dr. Unger and
Mr. Hugo Richter, of Davos, who are now living memorials
of the potency of the glacial atmosphere, which they inhaled
for the first time in 1865, at the instance of Dr. Spengler,
who was then country physician in Davos.

The first Englishman who visited the place was Mr. A. W.
Waters, F.G.S., in 1869. In 1871 he wrote " Klimatolo-

gische Notizen über den Winter in Hochgebirge," since which publication he has made many valuable and important contributions to the scientific literature of the mountain climate.

As may be expected, there are numerous sheltered spots in Switzerland where health may be sought with advantage, and where the benefits of living amongst a scanty population at a high altitude are associated with clean air and sunlight. The statement that the low temperatures of 5° and 10° Fahr., on sunny calm days, give an impression quite the reverse of the pinching cold of the British climate, flavours of exaggeration, but is nevertheless true. Frequently with the thermometer ranging between 20° and 30° Fahr., on still days, the dazzling spectacle, may without stretch of imagination, be characterised as a summer of snow ; but it must not be taken for granted that this state of the atmosphere remains constant from the commencement of November to the end of March. Anyone who gives a thought for an instant on the subject of climate, will remember that genial weather has never an uninterrupted run anywhere in Europe, whatever it may be in such places as Quito, Santa Fé da Bogota, &c., and some parts of California and Colorado, where the climate has been figuratively described by some as one of eternal spring.

With suitable clothing, however, the bracing cold air of Alpine winter is free from all dangerous chilling qualities, for the quantity of moisture suspended in the air is less than half that observed at Kew, whilst the heating power of the sun's rays is very great. Principally from these two characteristics, the occasionally low temperatures—even at zero— will bear no comparison with ordinary winter in the British Isles, where the solar rays are impeded and their heat absorbed by a moist atmosphere, the latter acting also as a more complete conductor of heat from the skin, than the dry Alpine air of the Grisons. The sensations, too, that are produced by the vivid beauty of the scene, the purity in colour of the blue sky in contrast with the overpowering whiteness and brilliance of the snow, are unknown to the stay-at-home Briton.

"Catching cold" is a rarity. Indiscreet exposure with unsuitable clothing may certainly lead to catarrh, sore-throat, or an aggravation of old symptoms; but in the majority of instances sore-throat, and colds, are contracted within doors from over-heated and stuffy rooms, cold corridors and cabinets, sleeping with the head of the bed against the external wall, dressing in a cold bed-room, taking insufficient exercise, sitting round a stove, or gazing at an open fire.

The fanciful idea of an open fire-place lies rooted in the English breast; but it is probable that nothing favours a cold in the head so much as watching a blazing fire; the face becomes heated to an unnatural temperature, and susceptibility to chill, with the dilated blood-vessels and open pores of the skin, is at its height.

The damp frigidity of bed-rooms in England is also a fruitful source of chills. These points are mentioned, as open fires are rarely used in the Alps for heating purposes, being quite unsuitable, and too extravagant in the matter of fuel; and it is a point well worthy of consideration whether the even temperatures found in the hotels which are uniformly warmed thoughout, is not one of the conditions which act favourably on diseased or delicate lungs, by rendering the patient less liable to chills and catarrhs.

It is essential that all persons who visit the mountains for health, or who desire comfort, should wear woollen materials next the skin, not only over the body, but covering the extremities as well. If clothed in this way it is astonishing what little difference is felt on leaving the house for a walk in the snow.

Another question of equal importance is the amount of exercise which should be indulged in by new comers. The counsel given to keep in the open air as much as possible, may if incautiously followed out on arrival, abridge some of the advantages to be gained during the first month's sojourn. This period is often unsettled as regards the weather, although the visit may not commence until the middle of October, and there is generally too much eagerness displayed by the novice to rush up the mountain sides, or take prolonged exertion.

This, if it does not lead to anything more than fatigue or loss of appetite, will be a bad economy of time, for the rate of improvement will certainly be retarded.

Putting aside unsound conditions of the lung, prominence must be given to the consideration of these climates for the *prevention of consumption.* There are hundreds of individuals amongst the young and middle-aged of both sexes on the verge of lung troubles, in whom an attack of simple catarrh or congestion would light up a life-long ill that might cramp many of the sweets of existence. Some, on the other hand, with undeveloped lungs imperfectly expanded in some corner, and who cannot take any hasty exertion without breathlessness; others losing flesh or becoming anæmic, with a family history not very satisfactory as regards lung vigour, or one gradually breaking down from confinement and mental strain.

It is said these people do well anywhere with rest and change. That is true; but the question at issue is, whether expansion of the chest and rapidity of improvement by mountain air is ever equalled even by a sea voyage. The Alpine pleasures of skating, sleighing, and coasting, are to many, far more enticing and health-giving than other sports one can mention, and by themselves roughly suggest to the mind the class of cases which would reap the largest benefit by winter out-door life; but whilst the highest altitude is repeatedly aimed at, observation suggests to me that there are many pulmonary cases who should begin their lung-training lower down the mountains than is commonly supposed. It has yet to be determined if the altitude of the Engadine is, or is not, too great to begin with, to obtain the fullest benefit of climatic treatment in every instance. Then the kind of life most suited to the individual who seeks change has to be taken into account.

Besides this, the selection of an hotel from a sanitary point of view must not be overlooked.

In suitable cases, judicious exercise and the special sports indulged in at these elevations, become natural therapeutic agents of great power on the development of the chest, and on morbid states of the lung. No drugs in the whole Pharmacopœia produce any effects resembling the inflation of an unexercised apex of a lung, or the expansion of an indolent thorax, in comparison with "coasting" and "tobogganing," or climbing a mountain side, rational remedies if used rationally, but dangerous if abused.

Those unacquainted with these sports may glean an idea of them from a short description ; but can never truly appreciate the exhilarating exercise except from personal experience.

"Coasting" is the term applied to runs made with a small iron-shod sledge over the polished roads, where the snow has been packed and pressed down by the traffic.

Towards the latter part of winter, runs can be taken across country on the snow crust which forms during the month of February. In the Bergell Valley some miles of country can be traversed if the winter is favourable to the formation of a crust and the snow covering deep enough to bury the rocks and short shrubs which would otherwise impede a long course. The sledge is guided by short sticks held in the hands, or a too rapid speed arrested by the feet. To the iron bar of each sledge is attached a small goat-bell, whose musical tinkle signals the gliding approach of the rider.

The "toboggan" is of another form, being used for sliding down inclines where the snow is not firm enough to bear the narrow runners of the sled. It may be described as a light thin board of elastic wood (generally made in two pieces of

Canadian elm) scrolled up at one end. Tobogganing is especially a Canadian sport, seen in all its turbulence on the huge ice cone which forms each year on the river St. Lawrence, at the base of the Montmorenci Falls. Three or more persons take their places on this wooden structure and start impetuously down the swift incline, whilst one of the number steers with a foot stretched out at the rear. Incautious dives down the surface of the ice-cone are by no means free from the danger of collisions and violent capsizes. In Switzerland there are no furious rushes from a pinnacle of ice, but the tyro is soon brought under the spell of the fascinating career down a precipitous descent, and quickly acquires the skill for manœuvering the nimble sled at a flying pace clear of all obstacles. The speed attained takes the rider on some distance over the snowfields below. These rushes through the air are fit only for the vigorous, weaker vessels attack a modest slope, and are content with milder capers than the plunges down a steep declivity sometimes entail, serious accidents are, however, rare ; a bad fall in the snow seldom does more than graze the skin. By far the most enjoyable runs are the excursions undertaken in glowing sunshine on the frozen roadways to some place of interest, or trips merely for the pleasure of a picnic, the return journey being made by sleigh.

It has been announced as a drawback to the climate of the Alps that the " cure " is not a permanent one, only a fair state of health being maintained as long as the patient dwells at these heights, and that he cannot live anywhere else. This is not improbable in a few instances, but such reasoning cannot be regarded, save as a demonstration, of the exceptional peculiarity of a climate that admits of life being prolonged. Whether an existence among the snows of the mountains is to be a happy and useful lifetime depends greatly on the habits, temperament, and energy of the individual, features which are quite distinct from the medical aspects of the case ; indeed, few persons with damaged lungs or other organs of the body, claim that their life shall be unconditionally extended by any brief special treatment.

Now, with all the peculiar beauty of Alpine winter climate, its exciting and stimulating air, bracing amusements and surroundings and bright sunshine, it must not be forgotten that such potent climates are not of universal application, and a word of warning is needed. Some invalids arrive on their own responsibility, others in direct opposition to medical counsel, which, to say the least, is a harebrained proceeding, and only brings discredit on the whole system of Alpine treatment.

DAVOS PLATZ.

ALPINE WINTER STATIONS.

CHAPTER I.

Davos Platz, Wiesen, St. Moritz, Pontresina, Samaden and Campfèr.

THE district of Davos (an elevated valley of the Canton des Grisons, in Switzerland) extends to about fourteen miles in length, and contains between 3,000 and 4,000 inhabitants. The three miles of valley, with the health resorts Davos Platz and Dörfli, has an altitude of 5,100 feet, and is surrounded by Alps ranging to 9,000 and 10,000 feet above sea-level. This portion of the valley is effectually sheltered from the north and west winds; rather less so from southerly currents. The larger village, Platz, is situated on the north-western side of the valley, and consists of hotels, "pensions," and shops. Of these there are four or five large hotels, seven or eight smaller ones, and several houses affording fair accommodation.

Some years ago the place was frequented principally by Germans and Swiss, but in recent times many English and French have wintered there, also Belgians, Russians, Spaniards, Dutch, and even Americans.

The English doctor established at Davos is Dr. William Huggard, who remains there winter and summer.

There are besides, several medical men in Davos or all nationalities, many of them have been in practice for some years, and have endeavoured in every way to sustain the repute of the little town—fathered as a *Kurort*, by the two medical gentlemen, Drs. Spengler and Unger.

The scenery appears picturesque and grand to a stranger in Switzerland, although somewhat limited in extent, the breadth of the valley being from 500 to 1,000 yards across. Numerous pine trees cover the steep slopes, and a winding stream, the "Landwasser," the greater part of which has undergone canalisation, receives the overflow from a lake situated to the N.E. of Davos Dörfli, is fed by numerous mountain rivulets, and acts as the main recipient for the drainage of Davos Platz.

The soil is dry and thin, excepting the central portion or bed of the valley, where it is of a peaty character. Nearly the whole extent of land is used for cattle-grazing and hay, and the sides of the valley are quite free from surface moisture.

Lateral dales—Fluela, Dischma and Sertig—lead into the main floor of the Davos Valley, and afford pleasant walks and excursions when the snow is not too deep. Skating and "coasting" are the prime amusements as with all the other Alpine winter resorts. Those who are not strong enough for these exertions, take walks, sit in the sun on balconies, and amuse themselves on their own resources. Crowded rooms with a smoky or stale atmosphere must of necessity be shunned if the winter visitor be in quest of health.

The old defective system of drainage has undergone alteration, none too soon, as within the last few years the increase in the winter population of Davos has been so large that this subject has forced itself on the notice of those most interested in attracting visitors to the place. It is a duty a successful health-resort owes to its *clientèle* to be liberal in any outlay relating to sanitation, the disregard of which will inevitably be a false economy in the end. A costly effort has been made in the present instance, that adds greatly to the attractiveness of Davos as a sanatorium, and reflects great credit on those who originated the scheme. As far as external drainage is concerned it cannot be denied that the town now stands amongst the first rank of health resorts, and in appearance presents a cleanly and favourable contrast to what it did formerly. A good deal of attention is paid to the roads in summer and winter, many paths are available for short or long walks, and what may be styled the metropolis of Alpine stations now assumes an aspect of neatness combined with a few not unmerited pretensions to some beauty. It is impossible for one who has known the locality for even a few years, not to be struck with the change the place has undergone. Hotels have been enlarged and added to; new ones built, and a few handsome villas erected.

A new and plentiful water supply has also been brought into the town from the Fluela Valley.

No one has at any time seriously made the attempt to uphold Davos as faultless; nevertheless, in the present instance it will serve a very convenient purpose to recognise in the well-known Alpine station a suitable standard for comparison. A great deal has been said for and against the place, both by those who understood something of the climate and by others who based their observations on short visits, or on accounts gleaned second-hand. In this way, although some error has crept in, much truth and information has been elicited, greatly to the benefit of those who had never seen the place, and who have eventually experienced the curative properties of its climate.

The reputation of the little town, although of many years standing, did not begin to grow with any very great rapidity until a well-known Englishman, now for some years a resident in Davos, visited the place for his own health. His experiences and friendship towards the valley have probably done more for its true appreciation by the English public than the efforts of any other person who has interested himself on mountain health resorts.

Some misunderstanding still prevails on the problem of "cold" as a sensation, many believing that the low temperature must necessarily be unbearable, and the thought of being surrounded with snow for the whole winter awakens visions of peril to invalids. There is little ground for any positive alarm on this head, as we shall presently see that low temperature on mountain heights does not produce the same effects of cold on the body that would be felt at sea-level. The results which have been obtained at Davos on patients with pulmonary complaints have been highly satisfactory, and some remarkable. Dr. C. Theodore Williams, who has published the details of many cases, has shown that the principal features noticed in persons who have undergone the Alpine cure are unusual expansion of the chest, gain in flesh, and improvement in sanguification. (a)

Notwithstanding the climatic virtues of the Davos Valley, its very popularity may eventually prove its bane. Already the smoke nuisance provokes unfavourable criticism ; also the state of the main thoroughfare—discoloured with the frozen contaminations resulting from the growing traffic, constant building is going on, and a railway to the place from Landquart is being constructed. To sustain the repute of the favourite resort a prudent and unselfish policy must be displayed by the local authorities, and support rather than opposition given to the development of new places, for it is well known that one of the chief points of a health resort for lung diseases is to have a scattered population, as well as plenty of cubic space indoors.

FRAUENKIRCH, 1½ miles from Davos Platz, has for

(a) "Pulmonary Consumption," 2nd edit.   By Dr. C. Theodore Williams.

WIESEN.

some years now received the overflow of visitors from Davos. There is very fair accommodation to be obtained there, and with the exception of the position being a little less calm than the upper portion of the valley, the climate is the same.

DAVOS DÖRFLI lies at the entrance to the Fluela Pass, about one mile from Platz, in the direction of the] Davoser See, is a few feet higher, and being situated opposite the Fluela Valley, receives more sun than Davos Platz. In other respects no great differences exist as to climate. There is, however, less smoke in the atmosphere of Dörfli.

Davos, like all other mountain resorts in the Grisons, can be reached by diligence from Coire, at which place a few days may be spent *en route* with benefit. Landquart, the railway station in advance of Coire, is nearer Davos, and if a quick journey is desired it can be made in about 32 hours by this route. In winter the post leaves Coire at 7 a.m., passes Lenz at noon, where a rest is made for dinner; at 1 p.m. arrives at Wiesen, and 4 p.m. Davos Platz. From Landquart the diligence leaves at 1.10 p.m., and, traversing the Prättigau Valley, gets into Davos at 8 o'clock, the distance being twenty-nine miles instead of thirty-four miles from Coire; but the Landwasser route from the latter place is well worth seeing, and the opportunity should not be missed of travelling over an interesting and wonderful road. The majority of persons, notwithstanding, take the direct route by Landquart.

WIESEN (4,771 *ft. above sea-level from Dufour's trigono-metrical measurements*).

This small village stands about twenty-four miles from Chûr (Coire), on the picturesque Landwasser road. About 11 miles further on, after presenting several structural difficulties, and passing through the Züge gorge, tho road winds into Davos.

The position of Wiesen exhibits many peculiarities and advantages as an Alpine health resort in winter. Situated on the slope of the Wiesener Alp, facing south and protected on the N. and E. by mountain ranges of 8,000 ft. and 10,000 ft. (Sandhubel, Foppa, Alteingrat, Leidbachhorn), it is effectually screened from the cold winds of winter. Sheltered equally on the southern and western aspect by the Buhlenhorn and huge Stulsergrat (8,390 ft.), the Tinzenhorn (10,278 ft.), and Piz St. Michel (10,374 ft.), and continuing the circle by the Motta Palusa, Piz Toisa, and Piz Curvèr (9,760 ft.), a land-locked region is formed, radiant in winter with dazzling sunbeams. A thousand feet below is the Landwasser chafing through a narrow course to join the Albula, a mile or so beyond.

Covering the slopes are innumerable pine trees. The odour from these is frequently perceived by new comers, and the antiseptic vapours exhaled presumably contribute to cure. The most noteworthy feature of the vicinity is an extensive plateau partially filling up the N.W. side of the basin, and jutting out towards the centre, forming an excellent promenade. The view from the border of this plateau is seldom excelled on a small scale. Far down at one's feet is seen the floor of the gorge, with the rushing Landwasser. A peep at the Albula Valley is gained, and beyond rises the Curvèr. On the left lies the little village of Jenisberg (5,010 ft.), the Stulsergrat towering above. Behind to the east is the Alteingrat, Leidbachhorn, &c., and on the right the two hotels separated from the plateau by a small vale down which the road descends to the Züge, passing the Känzeli waterfall.

A variety of attractive walks can be had in the neighbourhood. Limiting enumeration to a few within a short distance, the descent to the Jenisberg Brücke far exceeds all others in beauty. After repeated windings through a larch-wood a view is obtained of an undulating ravine densely overspread with dark green firs stretching away to the Albula Valley. The sound of the Landwasser becomes more distinct through the trees, and a few turns bring one abruptly on to the bridge. The bridge itself is built of pine, and has the sides and roof of the same material. It spans the chasm without any central support, and some rough engineering skill is evident in its construction. The spectator avails himself of a small hole, cut on either side, permitting a sight of the yawning abyss, 273 ft. beneath. It is the highest bridge in Switzerland, being 23 ft. higher than the Solis bridge, which spans the Albula, in the Schyn Pass. To project the head and peer down at the Landwasser is most fascinating, and well worth the half hour spent in making the journey. After leaving the bridge a few turns to the right will afford a fine front view of the Bärentritt. In this direction the return trip can be prolonged through the pines towards Ziegelhütte, returning to Wiesen by the Landwasser Road, occupying about one hour and a quarter.

Another walk which deserves special attention is to the Grosser Wasserfall. Passing through the village of Wiesen, distant about 200 yards from the hotel, and hidden from view by its position, the road makes several turns, and crossing one or two mountain torrents, reaches an open pasture. Traversing this and entering the forest, the waterfall is discovered in a secluded glade, affording a pretty glimpse at the peak of the Tinzenhorn.

The most frequented ramble, and one that has but a slight incline, is the Landwasser Road in Chûr direction, proceeding as far as the Tiefentobel. A commanding view is obtained of the Piz d'Aela (10,894 ft.) and the other lofty peaks on the south side of the valley. After passing over a stone bridge a tunnel cut through the rock is entered, and the road can be continued to Schmitten, about 2 miles.

By the same road in the opposite direction, towards

C

Davos, the Känzeli Waterfall is an attraction. A small gallery jutting out from the roadway has been constructed here as a point of view from which the pedestrian can see the torrent on the right, losing itself amongst the pines of a wild and rugged ravine. A continuation of the walk takes one on to the wonderful Züge gorge.

Distant 11 miles from Davos, 334 feet lower in altitude, and with a different contour of land, slight modifications in the meteorological details may be anticipated; whilst the general character of a cold, bracing, and stimulating climate is maintained. The chief points of variation from Davos are—its position on the side of a hill, sparser population, slightly higher and more equable temperature. In fact, a milder and less exciting climate than the higher altitudes. Wiesen may be considered to be about 2 or 3 degrees warmer than Davos, and, although an increase of mean temperature might lead to the belief that dangers would arise from thaws recurring, the periods of liability to actual snow-melting seldom take place with more frequency than in the latter valley.

With a higher mean temperature, of course the winter is shorter compared with Davos :—this might be two or three weeks at the commencement of cold weather, and about the same at the termination of the winter. During the worst time of snow-melting in Davos (which happens either in March or April, according to the mildness or coldness of the season), Wiesen clears of snow rapidly ; and there being no moist valley below, when the snow melts, injurious consequences, if any, proceeding from evaporation are avoided. The hard frosts at night during this period depend much on terrestrial radiation and gravitation of cold air ; the larger sky expanse presented to the Davos Valley favours the former, while the situation of the valley is low. The bad part of the winter is then prolonged, as the warmth of midday and afternoon causes a thaw : the evening and night change the wet snow into ice, and so delay its speedy disappearance. An early clearance of snow makes Wiesen a desirable locality for a change from Davos towards the end of winter. Or, by judiciously-timed transits from one place to the other much of the bad weather can be escaped at either station in the months of October and March. The journey can be made within two hours, and if a tolerably fine day be selected, little risk is incurred, even in an open sleigh. Covered sleighs may be chosen ; but with plenty of wraps, in bright weather an open vehicle is much to be preferred. Neither would it be at all a perilous test if the experiment were made of sending cases here when Davos was found to be unsuitable for them. If satisfactory progress were not induced, the patient would be spared the long drive to Chûr or Landquart, and would in any case only be breaking the journey. On the other hand, the advisability of making a rapid descent from a high altitude to the lowlands, even for convalescents is questionable. Many patients when leaving for the plains lose much of the ground they have gained in consequence, it may be, of the fatigues of travelling, exposure to windy or damp weather, neglect of ordinary precautions or care in personal management.

THE ENGADINE.—Leaving the Davos Valley with its four stations—Davos Dörfli, Davos Platz, Frauenkirch, and Wiesen—three passes are available for a journey to the Engadine,—viz., the Fluela (7,890 feet), Albula (7,582 feet), and the Julier (7,504 feet). Strange as it may seem to those ignorant of the winter climate of these regions, a passage fraught with but little risk can be undertaken in the depth of winter, and the journey to St. Moritz or the Maloja made in one day, even by persons who are not in robust health. Many change their *locale* in this way, and when tired of one place seek the novelty and freshness of new quarters. There seems no objection to the practice, always provided the wanderer is in a fit state of health for the trip.

. The Upper Engadine, or Upper Valley of the Inn, lies from S.W. to N.E., between the 46th and 47th parallels, is about twenty-five miles in length, and extends from the Maloja plateau (5,941 feet) down to the villages of Zuz (5,548 feet) and Scanfs (5,413 feet). The Lower Engadine, between Zuz and Martinsbruck (3,342 feet), although strikingly picturesque, does not equal in beauty the remarkable scenery between the Maloja and St. Moritz, with its magnicent chain of lakes, the largest and most beautiful being the Lake of Sils.

This is of all Europe the most elevated region that contains such a number of permanent inhabitants. Commencing at the Colline du Château (6,240 feet), Colline du Belvédere (6,084 feet), and Maloja Kulm (5,941 feet), a transverse ridge is formed, which slopes gradually down to the Maloja plateau. This ridge is perched over 1,200 feet above the valley of Bergell, into which huge chasm the descent can be made in a quarter of an hour, by a zigzag road having sixteen turnings. From the Belvédere this parapet bears resemblance to a gigantic balcony, commanding the approach from the plains of Lombardy, and sweeping the bed and sides of the vale, with a view rich in picturesque loveliness. A drive of two hours brings one to Promontogno; two miles farther is the Italian frontier of Castasegna, where flourishes a southern flora, together with the walnut, chestnut, mulberry trees, and vineyards. The little village of Castasegna is more than 3,600 feet below the level of Maloja. Another half hour takes us to Chiavenna, where is the lake of Como; whilst eighteen miles further south we have descended 2,000 feet.

From the Maloja, in the other direction, St. Moritz is reached by an excellent road in an hour and a half, passing *en route* the villages of Sils, Silvaplana, and Campfèr, bordering the adjacent lakes. 1st. The Lake of Maloja or Sils, resembling a miniature sea, facing the plateau, and extending over four miles to N.E. 2ndly. The Lake of Silvaplana and Campfèr, smaller in size, and divided by a projecting neck of land into two basins. 3rdly. The Lake of St. Moritz, the smallest of all, from whence the Inn descends by a rapid fall into the valley of Samaden and the Lower Engadine, making its way towards the Tyrol and Danube.

The valley of the Upper Engadine, about a mile wide at the Maloja, becomes a little narrowed towards St. Moritz, being enclosed by the Engadine Alps of the Albula range to the north-west, and to the south-east by the Bernina Chain, of which the peaks and glaciers count amongst the finest and most important of all Switzerland.

The landscape, taken as a whole, presents a panorama superlative in beauty, that leaves on each beholder an ineffaceable impression and lingering memory. The deep, still waters of the lakes, the emerald green pastures flourishing on their borders, their picturesque banks, present a striking contrast with the sombre green of the pine forests surrounding, and with those colossal masses of granite covered with eternal snow. In winter this scene is transformed by its snowy veil into resplendent whiteness, bespangled with myriads of crystals, whilst the sky retains its clearest blue, and the sun his summer power.

Cultivation of the land in the Engadine we may say is almost unknown. The dominating character of the country consists in its pastures, where the grass never attains a great height, but is green and nutritious. In summer, the meadows and slopes are covered with cattle, brought over by the shepherds of the Seriana and Brembana valleys, who leave again in August with their sheep in fine condition. The floor of the valley between St. Moritz and Maloja teems with a variety of gay-coloured little flowers, that one never sees at lower altitudes, and which adorn the swards with soft and gentle tints. Up to a certain height on the slopes of the mountains, are larches and red pines, and above all the *pinus cembra*, or cedar of the Alps, a species seldom met with in the forests of Switzerland, but flourishing in the south of Siberia to a height of more than 100 feet.

The flora of the Engadine is infinitely varied, and its richness to botanists can scarcely be surpassed. The Alpenrose (*rhododendron ferrugineum*) is found in abundance; the Alpine aster, (*a*) twenty species of the ranunculus, gentian, aconite, valerian, gnaphalium divicum, arnicum glaciale, &c., and lastly, edelweiss, which latter is only to be obtained at the price of a mountain ascent, frequently difficult and sometimes dangerous.

The air is so dry in the valley, that alone it is sufficient to dry meat, and to prevent putrefaction. It is also so clear and transparent, even during the warm weather of August, that objects at long distances can be seen clearly, and the sky is of an intense blue, such as one sees in Italy.

A broiling sun, insupportable during the summer time, at Venice, Rome, Naples, &c., each year directs towards the Engadine a ceaseless stream of emigration, from the rich and aristocratic families of the Peninsula, who do not come with the object of making a cure, but simply to escape the intense southern heat, and to refresh their frames with the bracing air of the mountain climate, where health can be renewed and the physical and mental faculties stimulated.

Situated nearly 1,000 feet higher than Davos Platz, the barometric pressure is lessened to about 3 lbs. on the square inch, instead of about 2½lbs. at the latter place. The winter season is longer, and a larger quantity of snow generally

(*a*) For the botany and zoology of the Engadine, *vide* Caviezel's guide.

ST. MORITZ-BAD AND DORF.

falls. The advantages of this cold and dry climate were represented in 1869, but at that period the high lands in Switzerland were not appreciated as they now are, and the experiment of sending patients failed through a misunderstanding of the requirements of such cases.

ST. MORITZ.—Dr. Symes Thompson has written an interesting brochure on winter residence at St. Moritz, embodying his experiences of the place, and of the progression of several pulmonary cases, which he had sent there with marked benefit.

In "Climate and Health Resorts," which gives a reliable account of nearly every European health resort of any importance, Dr. Burney Yeo furnishes interesting information regarding the birth and development of St. Moritz, both as a summer, and in recent years a winter residence. It would appear that when first the Engadine was tried as a residence for consumptives, an improper selection of cases retarded rather than promoted the appreciation of this altitude as a winter health resort, for the commencing seasons did not seem to succeed as far as pulmonary troubles were concerned. In the *Fortnightly Review* a very poor account was given of the results on some who spent the winter there. Doubtless the selection of cases referred to in the above article was a most unfortunate one, as the writer names six deaths as having occurred within his own knowledge. Of late years more certainty has been arrived at as regards the class of pulmonary cases which are likely to do well at these altitudes, and the results now, within the last six years, have been eminently satisfactory, so much so that St. Moritz is well started as a winter station for English visitors and invalids. The effort was made in 1882 by some gentlemen interested in preserving Davos from the consequences of too rapid growth, to re-develope St. Moritz for winter residence. This scheme was received favourably by two or three London physicians, and Herr Badrutt, the proprietor of some hotels there, eventually lent himself to the proposal. Dr. Holland then established himself at the Kulm Hotel, and about fifty persons were sent there by one London physician alone. Since that time the place has gradually grown in numbers, a newspaper, the *St. Moritz Post*, has been started by some of the more permanent English residents, and the social life in winter is enlivened by the usual in-door diversions which commend themselves to English tastes. The "Kulm" is a fine hotel, well-drained, with good bed-rooms and large salons, lighted by electricity and heated by steam. Other small hotels offer good accommodation to visitors, and are well filled in summer and winter. The village contains a few shops which supply all that is needful for short residence.

The village of St. Moritz (St. Moritz Dorf), seen on the rising ground in the illustration, lies in the Upper Engadine, about 11¼ miles from Maloja. Its highest point at St. Moritz Kulm, has an altitude of 6,089 feet above the level of the sea, and 285 feet higher than the shores of the lake. The village, which is the highest in the valley, is well sheltered from the north and east winds, and being elevated out of the bed of the valley escapes a good deal of the thalwind. It possesses all the characteristics of an Alpine resort—sunshine, calmness, low temperature, high solar radiation and altitude. It gained a reputation in summer through its famous chalybeate springs. In 1856 the Kurhaus was built by a joint-stock company, and in 1866 it was found necessary to enlarge this establishment.

The winter specialities of St. Moritz are the out-door sports and amusements undertaken by the more robust of the invalids and visitors. Interest is centred on toboggan runs, ice-rinks, and curling. The life at the Kulm has less of the town element than the life led at Davos. In spite of perhaps a little indiscretion here and there, the majority do well, and the results have been in the main highly gratifying. A somewhat pleasant winter existence is spent, even if the full benefit of a systematic "cure" has not been attained in every instance. There are many visitors too who are fond of skating and the delights of snow-life, who during the winter season spend several weeks at this lovely spot. They say the stimulating atmosphere makes them feel "fit;" and a summer holiday in the best months of the year in England is often curtailed or yielded up to the Engadine ice-god in the months of December and January. They have their reward, for nothing on the face of Europe can compare to the crystal brilliance of these days in mid-winter, with the refreshing and crisp mountain air.

In summer time there are several agreeable and pretty walks to be made from St. Moritz, as well as mountain ascents. In winter many of these can be undertaken in snow-shoes without much fatigue, but in all the frequented paths and highways walking may be done on foot without the inconvenience of the summer dust. One of the most pleasant promenades, which is much frequented, is in the direction of Cresta, descending the wooded slopes which meet the toboggan-run where it crosses the high-road. Three-quarters of a mile further on the village of Celerina is reached, and from thence to the right Pontresina (three miles). Samaden is on the same road, about the same distance from Celerina as Pontresina, and level walking instead of the slight ascent necessary to gain the latter place. A picturesque excursion, and one frequently made, is that leading towards Campfèr (2½ miles), where a good view of the upper portion of the Engadine with the lakes of Campfèr, Silvaplana, and Sils. The towering Piz Margna, which overlooks the Maloja Kursaal, is a prominent object, and is the peak which can be kept in sight down the Engadine as far as Zuz. Besides these expeditions rambles are made in side directions, and with the aid of the snow-shoe one is free to go almost everywhere.

On the road to Pontresina (by the borders of the St. Moritz lake) the little lake of Statz affords the first skating of the season, sometimes as early as November. If the weather is propitious as regards a snow-fall, some days sport may be had on the ice there ; but November is a snowy month, and a week's good skating, without having occasion to sweep the ice, would be considered an exceptional event.

PONTRESINA (5,915 feet).—Three miles east of St. Moritz, at the foot of the Bernina pass. A few visitors have wintered here, but it is mostly in the summer months that this attractive spot is frequented. The chief feature of Pontresina is its early sunlight. The village being placed on the slope, and having a magnificient mountainous but comparatively open view to the east and south, receives the first rays of the sun at 8.10 a.m., and 8.30 a.m., during December and January ; whilst the sunset varies in these months from 3.5 p.m., to 3.20 p.m. The air is bracing and dry ; the day-temperature is said to be about the same as Davos.

SAMADEN (5,600 feet).—The principal village in the Upper Engadine containing over 800 inhabitants and several shops. There are generally some visitors to be found at the Bernina Hotel in winter, they come mostly as a change from other places, but some remain throughout the season. Skating rinks are kept in order and level walks are available. This spot may be said to have entered a claim to be considered an Alpine station as well as Campfèr, situated above the Julier Pass, and fronting the meadows which face the lakes of Campfèr and Silvaplana. The splendid views of the valley of the Inn surround both these villages, and their climate varies but little from other parts of the Upper Engadine.

CAMPFÈR (6,000 feet).—This is a quiet little village, with fairly good hotel accommodation, and will, no doubt, in the course of time, become patronised for winter abode. There is plenty of flat ground for promenading, and usually the lake situated here freezes during an interval of the December snows.

It is almost with uniform regularity that the annual freezing of the Engadine lakes sets in at the St. Moritz end of the chain about the commencement of December, and extends gradually up the Campfèr and Silvaplana Lakes to the large Silser See (Maloja Lake), where it finishes off at the Maloja end of this vast expanse of water with singular periodicity during the week preceding Christmas Day.

Another place which has been set forth as a winter health resort is Andermatt (4,738 feet), situated in the Urseren Valley. From the unusual facilities offered by the St. Gothard line, this spot deserves to be noticed, being only 29 hours from London, with a drive of less than an hour for mounting from Göschenen. The position of this locality is excellent for convenience in travelling either to England or to the Riviera, if the mountain air is found to be unsuitable, but no personal experiences can be given here extending over sufficient time to warrant an opinion on its meteorology, or of its fitness for a winter station.

LES AVANTS (3,212 feet), is one of the half altitude resorts. For several seasons winter visitors have been attracted here. The place is well sheltered from north and north-east winds. The temperatures are not quite so low as the higher stations, but skating and tobogganing are indulged in.

One of the most picturesque little spots in the Grisons— "that mysteriously hidden wrinkle in the stern grey hills, with its fairy lakes, solemn pine-forests, and emerald meadows"—Arosa ; must not be omitted from notice. Strangers are known to have wintered there and reliable

accounts seem to point to its being the perfection of winter
climate.   In summer the few inns are filled with Swiss
fugitives from the heat of the plains near Coire, who preserve
the secret of the unassuming beauty and charming climate
of the concealed nook.   Arosa is reached by a bridle path
from Langwies, 2½ hours on foot.   The construction of a
road from Langwies or from Coire would soon develop
Arosa into a well-patronised abode.   One portion of the
valley stands at an altitude of 6,207 feet above the sea-level,
the whole area appears well-protected from wind, but as
regards sunlight in the depth of winter, no observations have
been noted.

CHAPTER II.

Water and Soil.—The water supply of the dwellings at
all the high level stations is upland surface water, and is of
wholesome and excellent quality. It falls on the extensive
summits and slopes of lofty mountains, as mist, rain, and
snow, and undergoes filtration in its descent to the valleys,
where it makes its appearance in small streams and springs.
The rivulets selected to supply the houses are protected from
cattle, &c., and the water conveyed in iron pipes to the
hotels ; sometimes wooden tubes made by boring through
pine trees endwise, are used as pipes. In winter it is constantly running to prevent the pipes freezing. The likelihood
of contamination is then at a minimum, as no storage takes
place in the houses, nor are there any inhabitants above the
level or in the vicinity of the mountain streams. The water
itself is without odour or flavour, and has a bright spark-
ling appearance from the carbonic acid taken up in the
interstices of the rocky soil, through which the springs
percolate. Although deprived of its oxygen in winter, by
freezing, when falling as snow, it becomes sufficiently aërated
in its course to the lower elevations. At Wiesen the water
is drawn from sandstone, which makes it considerably softer
than if obtained from the Dolomite limestone, as is generally

the case at these levels. An analysis has been made of the water at Davos Platz, by Philip Holland, F.C.S., Public Analyst for Southport, who found a somewhat considerable proportion of mangesium carbonate (4·506 gr. to the gallon), but considered the water to be exceptionally pure.

The soil of Swiss mountain pastures is thin and of a rich, fertile character, absorbing roughly 1·4 (a) times its weight of water, underneath is solid rock, mostly Dolomite lime-stone, except in the centre and at the edges of the valleys or ravines, where lie the moraines of ancient glaciers composed of rubbish or deep earth, intermixed with smooth blocks of rock and pebbles.

From the steep declivities on which moisture falls no injurious collection of water can take place. The drainage both of the surface and subsoil is rapid and effective. Being also of shallow depth, the danger arising from a heavy rain-fall pressing up effluvia from the deeper strata is obviated ; and having also no stagnant underground sheet of water, the evolution of organic emanations, forwarded by constant moisture, is not aided ; or if these decompositions do occur, it is probable that absorption of their deleterious properties, is mainly affected by contact with comparatively dry earth. Here and there small basins are formed in the irregularities of the rocky formation of the foot hills ; in these hollows a miniature bog is occasionally alighted on, but the danger to health from this source is insignificant. Vegetation also, as short thick grass covering the slopes, speedily absorbs the products of decomposition. The condition of the land with which we are principally concerned during the winter season, takes on altogether a changed aspect, and one which until recently has attracted little attention. · Uninterruptedly

(a) 2,500 grms. of soil taken from a garden at Wiesen and dried, weighed 1,850 grms.   On being wetted again and allowed to drain, weighed 2,600 grms.

covered with snow for about four months during the year, the ground itself assumes the nature of "sub-soil," whilst the layer of thick snow above the actual surface modifies in a marked degree the effects of telluric influence.

Atmosphere.—As the purity, freshness, and stimulating properties of the air are everywhere important points in climate, attention must be directed to its characters of humidity, temperature, barometric pressure, sunlight, wind, cloudiness and transparency.

The quantity of moisture in the atmosphere affecting the rate of evaporation from the lungs and skin is a point which requires much consideration; but at present observations have been rather limited, so that we have not much data to go upon. Certainly, warm moist air, is sometimes very grateful to cases of congestion and irritability of the bronchial tubes; but a more permanent condition of moisture, of which our English winter affords an example, undoubtedly has an injurious influence on the majority of lung diseases; the watery vapour impeding evaporation of waste products, and abstracting an undue amount of heat from the skin and respiratory tract, and giving rise to catarrhs, congestions, and inflammations. It is a generally accepted fact that dampness of soil, apart from hereditary tendency, favours the development of phthisis. This dampness supplies the atmosphere with large quantities of moisture keeping the lower stratum of air in an almost constant state of saturation. Now, it may be expected that in places where the surface formation is favourable for the rapid escape of surface water, where dry snow covers the ground, and the rainfall is slight, the reverse of this condition may be expected. Such is the case in the high altitudes of Switzerland; although an occasional thaw or fall of snow may saturate the air with moisture, the actual quantity of watery vapour held in suspension during winter time is extremely small, from the fact that air, at a low temperature can contain but little, and the lessened atmospheric pressure favours rapid evaporation.

What the precise pathological effects are, of the habitual breathing of moist air is not yet quite clear; but Miquel has clearly shown that microbes are not found to be so numerous in moist weather at the Montsouris Observatory, Paris, as during dry weather; nor are they present at any time in the actual vapour of water.

As regards phthisis, however, Haviland's map indicates plainly that the distribution of that malady is decidedly in favour of damp places. (a) Excessive cooling of a part of the respiratory tract from moisture, on account of the conduction of heat being more free (b) in a damp atmosphere, is frequently experienced when breathing a cold London fog, the sensation of cold being felt sometimes far down the air passages. The result on the skin is to abstract an undue amount of heat, and retard the normal evaporation of waste products, thereby throwing more work on the lungs and kidneys. Although there is a diminished death-rate from phthisis in some parts of Scotland and Ireland, where excessive moisture prevails, the people there are mostly employed in out-door pursuits, which makes up for a great

(a) During the progress of cutting the St. Gothard Tunnel the workmen employed were exposed to a *high* temperature (80° to 90° F.) and an excessively damp atmosphere. Their health in consequence suffered severely, one of the principal complaints being accompanied by an intestinal parasite. Catarrhs with unusual dyspnœa were constant, and the colour and appearance of the men indicated imperfect aëration, of the blood, and consequent anmœia. Their strength and appetite failed them, and the body temperature was raised. There was great mortality also amongst the horses used within the tunnel.

(b) It takes a great deal more heat to raise the temperature of watery vapour than that of dry air, the apparent specific heat of which is, according to Regnault, ·2379 with pressure constant. Specific heat of vapour of water ·4750. Water 1·000, all at 32 deg. Fahr.

deal in the matter of health; but if a phthisical patient courted extreme ventilation in a damp climate the result would not be so satisfactory as exposure to drier surroundings. Perhaps it is necessary to mention that air containing less than 25 per cent. of moisture, with a temperature of 60° Fahr., is too dry for health, and might give rise to irritation of the air passages.

It is remarkable what misunderstanding prevails on the subject of "humidity" in relation to "equability" of climate. How many medical men recommend a "dry and equable" climate for lung affections? Books are written extolling certain places as being "dry and equable," and one of the first questions of invalids is to inquire if a certain health resort is dry, and limited in range of temperature. It is not too much to say that there is no climate on the face of the earth which is "dry and equable." Nearly every climate which has a small range of temperature is necessarily a moist one, for it is the alternate evaporation and condensation of moisture which limits the range of the thermometer. As soon as temperature rises increased evaporation cools the air, as soon as temperature falls to the dew-point, condensation of moisture raises the temperature by liberating the latent heat in the condensed vapour. It is in this way "equability" is maintained, and the various communications published, announcing this or that locality as being favoured with a very limited range of temperature, indicate simply that it is a damp climate. Even California, which is perhaps the driest of equable climates, is kept equable by an ocean wind, which must necessarily be more or less moist.

Denison, in a brochure on moisture and dryness, written in 1885, adverts to this and says. Hitherto it has not been customary for climatologists to see any good in variability.

If it is attempted to make an exception in favour of such places as Cape Mendocino or other California towns bordering on the ocean, let it be understood that though the ground

D

is dry and parched, the equability is due to a continuous moist wind, which blows landward from the Pacific Ocean, and does not part with its equable characteristics until, rising on the interior ranges of mountains, it has lost its moisture and become dry, and therefore variable.

On the other hand the dry climate of South Africa is well known to be variable in temperature, and the heights of Colorado and New Mexico, the driest climates in the world, are liable to unparalleled variations in the thermometer. It will be seen further on that these variations in a dry atmosphere are not so objectionable as is generally supposed.

The localities under consideration are shut in by high mountains, glaciers, and snow fields ; and however laden with organisms, currents of air may be, distant winds are partially or wholly sterilised by becoming attenuated and cooled to a low temperature in their passage towards these valleys.

Dust.—In the elevated lands of Switzerland the snow buries all impurities, and noxious emanations are destroyed or lie dormant until the spring thaw. Frequent falls of snow during the winter keep any surface refuse, as it were, below ground, and assist to free the atmosphere also, from aërial germs and mechanical irritants, fixing them in a freezing medium ; the former being destroyed or absorbed by the soil and vegetation, at the time of thaw ; the latter being washed into, and mixed with the earth. Under these circumstances it is apparent that the air above is as pure as Nature can produce it—clean, dry, and laden with balsamic vapours from the pines. There is almost complete freedom from dust, with the exception of the carbon, &c., from the chimneys, which, of course, depends on the number of habitations, and the stillness of the air.

Therapeutical Effects of Cold.— Moderate cold increases the appetite from its exhilarating and stimulating powers, and from the requirements aroused in the body for more hydro-carbons to meet a normal increase in combustion. A

well-known writer on therapeutics remarks that the most
vigorous health is maintained by a rapid construction and
destruction of tissue within certain bounds, provided these
processes are fairly balanced. Cold, when judiciously applied,
is well known to be a powerful tonic. A cold climate and
cold bathing are tonic and bracing. The theory of the tonic
action of cold may, perhaps, be stated thus :—During exposure
to cold the loss of body-heat, as tested by the thermometer, is
by no means the measure of the quantity withdrawn. Many
observers have shown that at such times increased combustion
occurs, whereby much of the lost heat is compensated, and
the temperature is maintained as soon as restored. This
increased oxidation of tissue is demonstrated by the greatly
increased quantity of carbonic acid thrown off by the lungs
on exposure to cold (Ringer). Moderate cold may be considered a nervine tonic, as it stimulates the nervous system,
dispelling the languor and want of energy that heat produces ;
the most striking example is the " plunge " or " cold douche "
after a Turkish bath, and also the capacity for exertion that
one possesses in a cold climate. With regard to sanguification,
a low temperature causes the lungs to absorb more oxygen,
and by thus inducing quicker change in the blood corpuscles,
exerts a highly beneficial influence on anæmia.

Many patients lose their night sweats and high temperatures in the Alps—sometimes after a few days' residence.
The lower temperature of the surrounding atmosphere, provided they keep their rooms cool, influences this to some
extent, and also, in the case of perspiration, its stimulating
properties brace up the coats of the cutaneous vessels, and
aided by the diminished barometric pressure, promote a
healthy action of the skin and sudoriparous glands. Some
analogy appears to exist between the action of cold on
temperature, and the effects of quinine and salicylic acid :
these agents often reduce a high temperature, but have a
markedly diminished action on the normal heat of the body.

Moderate cold appears to influence the desire, as well as the capability, for movement and exercise ; thereby maintaining the body in working order or training deficient organs up to a state of activity by the stir which Nature urges, for the comfort, satisfaction, or warmth of the individual. It can be readily understood that the employment of the various organs and limbs of the human frame is essential to their efficiency. From disuse, man has lost the grasping power of the foot, and the toes of the present species are atrophied, although the tendons remain. It becomes a question to be considered whether the sedentary occupations of modern civilisation will not eventually favour a pseudo-atrophy of organs connected with the respiratory and circulatory systems. How many men, dwellers in any of the large cities, having reached the term of years generally styled the "prime of life," can run a hundred yards with comfort, or mount a staircase rapidly without breathlessness and palpitation ? With women the case is even worse, an additional curb being placed on the act of respiration by unyielding corsets.

Rattray's observations on the weight and height of forty-eight naval cadets, aged from $14\frac{1}{2}$ to 17 years, during four successive changes of climate on a voyage, show that in the Tropics they increased in height more rapidly than in cold climates, but that they lost weight very considerably, and, in spite of their rapid growth, Rattray concludes that the heat impaired the strength, weight, and health of these lads. His figures seem conclusive on this point, and show the beneficial influence of cold on youths belonging to races long resident in temperate climates. (Parkes' " Hygiene," p. 436.)

Sunlight.—There is great difficulty in estimating the exact effects of sunlight on the body, but there are many reasons for supposing that solar rays have an important bearing on the health and condition of the body, when not associated with too high a temperature. The sun, as we know, is the central body and luminary of the solar system, and the source of all life on this planet. Without the light and heat given to us from this orb, all animal and vegetable life, except the lowest germs, would become extinct. We may fairly assume then that the light which is given off in such intensity in the clear regions of the earth has an important result on the individual exposed to its rays. The exhilarating and mental effect alone is cheering, and acts on the nerves in a way not to be explained, but experienced by everyone who has undergone the sunning process in a cold climate, where picnics and luncheons can be taken with comfort in the open air, whether it be on the icebound lakes of Nova Scotia and Canada, or on the snow slopes of the Alps.

It may be a mere coincidence that rabbits, mice, and some animals which get less sunlight than others, are extremely susceptible to the poison of tubercle, and are apparently unable to throw it off, unless exposed to plenty of light and air.

The enormous amount of light which is reflected on to the individual inhabiting snow lands, and which necessitates the wearing of black glasses to preserve the eyes, and sunshades to keep off the heat, is a factor in climate that has been under-estimated. It can easily be proved by the aid of a sensitive photographic plate that the sun's rays penetrate clothing.

That light has a favourable action on the blood corpuscles may be easily seen by observing the large number of workers in mines and dark factories, shop girls, clerks, &c., who suffer from anæmia ; likewise stokers on board ship and sailors employed entirely between decks or in the hold of a vessel, where the amount of light is necessarily limited, or perhaps entirely absent ; these men, if contrasted with the workers on the upper deck, compare unfavourably in healthy appearances, although as regards diet and sea-air, they are very nearly under the same conditions. Referring to Arctic experiences, there is every reason to expect the converse of the depressing influence exercised by the prolonged and intense darkness of the Arctic night.

Dr. W. Hammond contributed a paper in the "Medico-Chirurgical Transactions," on the influence of light, showing that the development of tadpoles may be retarded by depriving them of light ; and that in an experiment with two kittens, where one was confined in a dark box and the other in a box to which light was admitted, the weight was perceptibly increased by light, while the growth of the other was retarded. Various experiments demonstrate that the action of light is of benefit in many conditions, especially anæmia and phthisis. Other factors enter largely into the cause of anæmia, but the want of sunlight bears on it very strongly. The aspect of health which is created by the sun's rays speaks for itself, showing that light is a therapeutic agent of much value.

The chemical action in plants depends greatly on the presence of sunlight with chlorophyl, and although green plants break up carbonic acid and liberate oxygen, they do the reverse in the dark. Some gases also which do not combine in darkness immediately do so on exposure to strong light.

Barometric Pressure.—In approaching this subject it will be interesting to first quote the experiences of Glaisher, Gay Lussac, and others, of its result on the action of the pulse and respiration.

Balloon ascents of		Feet.	Increase in pulse.
Biot and Gay Lussac	.	. 9,000	. 18 to 30
Glaisher.	. .	. 17,000	. 10 to 24
		24,000	. 24 to 31

An ascent by Glaisher and Coxwell on the 17th July, 1862, gave these results :

Mr. Glaisher's pulse . . . 76
Mr. Coxwell's pulse . . . 74
At 17,000 feet, Glaisher . 100
    ,,   ,, Coxwell . . 84

21st August at 1,000 feet.		At 11,000
Mr. Coxwell	. 95	. 90
Mr. Ingelow	. 80	. 100
Captain Percival	. 90	. 88

The humidity of the air was found to decrease with the height in a wonderfully decreasing ratio, till at heights exceeding five miles the amount of aqueous vapour in the atmosphere was found to be very small indeed. (a)

The number of pulsations usually increased with elevation, as also the number of respirations. (b)

Armieux, in the case of eighty-six invalids removed from the plains to Barèges at a height of 4,000 feet, satisfied himself, after a residence of four months, the respirations were increased by two, and the beats of the pulse reduced by four. He had also found on careful examination that the eighty-six men had in four months gained on an average one inch in girth round the chest.

Dr. Kellett found that the invalids at Landour gained one inch, chiefly during the first two weeks. (c)

Jourdanet has asserted ("Le Mexique") that the usual notion that the respirations are augmented in number in the inhabitants of high lands is "completely erroneous: that the respirations are, in fact, lessened; and that from time to time a deeper inspiration is involuntarily made as partial compensation."

However one may disagree with this writer in some of the theories which he advances, one cannot but appreciate

(a) "Lectures in Exeter Hall," by Glaisher.
(b) Glaisher's "Travels in the Air," 1871.
(c) "M. R.," vol. lviii., 1876.

the valuable contribution he has made to the literature of
high altitudes by his work on " Mexico and Tropical America,"
1864.

Coindet from 1,500 observations on French and Mexicans
does not confirm Jourdanet's statement. The mean number
of respirations was—

<blockquote>
19·36 per minute for the French,
20·297   „   „   Mexicans.
</blockquote>

Evidence is in favour of a slight increase both in the
pulse and respirations in persons first dwelling at high altitudes, but the length of time these phenomena last has not
been noted with much accuracy. It must not be forgotten,
however, that the increase in the measurements of the chest,
and the excursions forward of the sternum, after a short residence in Alpine valleys, may also in a great measure be
effected by the gain in flesh and strength, for we know that
on recovery from many diseases at *low* altitudes this event
is a consequence of returning health, and a token of general
convalescence. As far as my own observation go, on patients
with delicate lungs, I find the following conditions on
arrival at 6,000 feet :—

1st. Increased pulse rate, the number of beats depending a good deal on the activity and extent of pulmonary
disease.

2nd. Increased number of respirations in all persons
with inactive skins, who do not perspire readily.

3rd. Diminished number of respirations in some cases of
phthisis. These are usually the sweating cases.

4th. The pulse rate and respirations are never diminished in health, generally increased, the latter with sighing
breathing at intervals.

In favourable instances the pulse as well as the body
temperature begins to drop during the first fortnight. The
first change to observe in the action of the heart is the difference in the number of pulse beats when in a sitting posture

and in a standing position, after walking across a room. I have frequently found the variation in the two positions reach as much as 20, 22, and 24 beats, without being accompanied by any unusual dyspnœa or palpitation. As strength increases these numbers fall gradually to a difference only of 4 or 8 per minute, in the altered position of the body ; and preceding improvement of the physical signs of disease, the pulse drop in number and decrease in tension, without much alteration in the breathing powers.

With very few exceptions, the chest expands even after a month's residence, that is, with patients who are able to take a fair amount of exercise, but I have never observed the amount of expansion named by Dr. Kellett, as having been noticed by himself in the Himalayas at 7,300 feet.

Oxygen.—The difference in the amount of oxygen inhaled at an ascent of 6,000 feet is as follows :—

In a cubic foot of dry air at freezing point and 30 inches barometric pressure, we find 130,375 grains of oxygen. A man draws on an average when tranquil 16·6 cubic feet of air into his lungs per hour, $130·375 \times 16·6 = 2164·2$ grains of oxygen (Parkes). An ascent (about 6,000 feet) where the barometer stands at 24 inches, will reduce this 1-5th, or $\left(\frac{21 \times 130·4}{30}\right) = 104·32$ grains, lessening the quantity per hour by 432 grains.

Without allowing for a slight difference of oxygen in favour of high altitudes, owing to the small amount of moisture in the air, and its increased density by low temperature (making a difference of 120 grains per hour between 0° and 15° centigrade) three additional respirations per minute would be necessary to compensate for a barometric fall of six inches, provided that all the oxygen of respired air was utilised in the lungs, which is not the case. Practically such an increase in the number of respirations never takes place. The physical effect of lessened pressure exerts its force

chiefly on the heart and circulation, and increased respiratory
movements are noticed on exertion, as being out of all proportion to the same exercise at sea-level. This indicates an
excess of blood in the pulmonary circuit. By experiments on
animals it has been found that as long as the percentage of
oxygen was not below 14, the same quantity was absorbed
into the blood as when the gas was in the normal proportion.
The quantity of oxygen in the atmosphere surrounding animals
appears to have very little influence on the amount of this
gas absorbed by them, for the quantity consumed is not
greater, even though an excess of oxygen be added to the
atmosphere experimented with (Regnault and Rieset). It
therefore does not seem at all probable that the lessened
weight of oxygen taken into the lungs, when breathing rare-
fied air at 6,000 feet, necessitates any increase in the number
of respirations. This is not unworthy of notice when we
reflect that in phthisis and some forms of anæmia there is
diminished respiratory function. It is not that the whole
quantity of oxygen taken up by the blood (supposed to be
held by the red corpuscles) is increased at great heights, on
the contrary it is lessened by reason of diminished pressure,
but that *chemical action* in the combination of this oxygen
with the carbon of the body is aided by diminished pressure,
and although less oxygen is carried by the red corpuscles, it
is more freely given up to form carbonic acid.

The explanation why breathing mountain air should
expand the lungs cannot be satisfactory accounted for by the
laws of mechanical pressure, as the diminution is everywhere
the same, both internally and externally, and such an equili-
brium of force being established disposes of any theory which
attributes increased thoracic capacity, directly to diminished
barometric pressure.

The rhythm of the involuntary movements of the chest-
walls and diaphragm depend entirely on nervous influence,
and it would appear that the cause of extended respiratory

movements depends on the excitation of the respiratory centres, influenced, amongst other causes, by certain fibres which run in the course of the pneumogastric. Rarified air irritating these fibres, and causing a desire to breathe a denser atmosphere, would account for the additional number of respirations and extended chest movements. An increased proportion of blood circulating in the lungs would also tend to this result.

That the lungs contain more blood in a cold climate is pretty clear if we accept the evidence of Dr. Francis (Bengal Army), who found from a large number of observation, that the lungs are lighter after death in Europeans in India than the European standard. Parkes confirms this, and also Rattray, in his observations of diminished respiratory function in hot climates.

In the cold high altitudes we may then, attribute the change to both these causes, but whether this phenomenon occurs in persons whose lungs are already quickened in action by disease is still a matter for further observation. In these cases the exciting cause of respiratory rhythm depends more on the proportion of carbonic acid and oxygen in the blood than on the density of the air breathed.

The following facts prove that this condition of the blood influences the respiratory movements :—1. The respiratory movements can be totally arrested if, either by a forced artificial respiration (by blowing air into the lungs) or by forced voluntary breathing, the blood becomes saturated with oxygen and poor in carbonic acid ("apnœa"). 2. Respiration becomes stronger, and the more accessory muscles take part in it ("dyspnœa"), the poorer in oxygen and the richer in carbonic acid the blood is ; as, e.g., on the entrance of air or fluid into the pleural cavities, causing a collapse of the lung, or when, by inflammation, &c., the lungs are unfit for respiration (Herman's " Physiology ").

In impaired lungs with quickened action (owing to

excess of carbonic acid in the blood), the increased elimination of CO_2 at high levels seems to allow of an inhibitory
influence.   This appears to be borne out by the fact of
some patients breathing comparatively freely when leaving
the plains who again experience dyspnœa on their return
from the mountains.   The sudden exhilaration of spirits produced by the mountain air, often felt on the journey up,
points rather in this direction, as well as the improved colour
of the skin and mucous membrane seen later on.   Muscular
exertion, too, is felt to be less fatiguing (muscular fibre is
stimulated by oxygen).   From the experiments of Dr. Marcet
at high altitudes in Switzerland and the Island of Teneriffe,
it appears that more air in bulk, but less in weight, is
breathed at high altitudes, and that a larger proportion of
carbonic acid is excreted in the cold altitudes of Switzerland
(showing that more oxygen has been absorbed).   At Teneriffe
the carbonic acid was not increased in amount, whereas in the
Swiss altitudes of 13,000 feet an increase of 15 per cent. was
discovered.   We can attribute to the lower temperature of
the latter country (as Dr. Marcet observes), and also to the
larger amount of sunlight, and in a *cold* high climate there
would be a greater volume of blood in the lungs to avail
itself of the absorption of the oxygen, and to excrete carbonic
acid.   A small amount of CO_2, about 1-50th of that excreted
by the lungs, is thrown off by the skin, and oxygen absorbed
in proportion, but in all probability the respiratory function
of the skin is more considerable in the higher animals than
is supposed.   Whether this is too insignificant to be regarded
or not, there is scarcely a doubt that the stimulating or sedative character of a climate commences by its effect on the
sympathetic nervous system, and in the chemical action of
the surface of the body as well as in the lungs, blood, and
tissues.   An atmospheric bath in one locality—where there
is sunlight, dry air, ozone, high electric tension, volatile
products from vegetation, and certain emanations from the

soil—must have a different effect, not only on account of the air respired, but by contact with the exterior of the human body, than an aërial medium of another composition, (a)

One may be said to live quickly at these altitudes, and the most perfect health is maintained by a rapid waste and repair of the tissues of the body. This increased combustion does not mean shortened existence, but improved health, provided that repair and loss are equally balanced.

Before leaving this subject, it will be interesting to note the diminution of barometric pressure, in the Engadine, viz., 677,476 lbs. off the whole of the human body. Although the considerable reduction of three tons (b) taken off the fluids, and solid parts of the skeleton carries with it no remarkable phenomenon, it is not illogical to assume that such a declension in the weight of the atmosphere must exert some peculiar powers. Whether these affect physiological action or influence the animal economy in any way, other than favouring chemical action; increasing the beats of the pulse and the number of respirations; rendering muscular exertion less fatiguing; may be elucidated by future observations.

Dr. Henderson, who spent a whole winter at Maloja made some sphygmographic tracings which tend to bear out the theory that less force is required at high levels to propel the

---

(a) There is evidence that the interchange of gases between the air and the blood through the skin has an important share in keeping up the temperature of the body, and we find the temperature of the surface much elevated in cases of pneumonia, phthisis, &c., in which the lungs seem to perform their function very insufficiently (Carpenter's " Human Physiology ").

(b) The barometer is lowered about 6 inches, 1·5th of the capacity of the mercurial column at sea-level. The cubic inch of mercury weighs 3,433·5 grains, or ·49lb. The surface of an ordinary sized man is about 16 square feet : therefore the calculation can be easily made.

blood through the vessels. (a)   This, he goes on to say, diminishes the work of the ventricles.

"*Secondly*—The superficial blood-vessels will be more affected by the loss of pressure than those deep-seated; consequently the surface of the body, as well as the lungs, will be better supplied with blood.

"It is quite conceivable that this disturbance in the distribution of the blood might be *completely* corrected by vaso-motor adjustments, but an examination of the sphygmograms annexed will, I think, prove that in point of fact this does not take place.

"The sphygmograms, exhibit in a marked degree the signs of loss of tension—striking development of the percussion wave, great breadth of wave, great breadth of trace, with fall of the base line.   I think it will be admitted that the loss of tension is due to increased freedom of circulation in the capillaries from relaxation of their walls, and that this denotes a larger amount of blood in the part.   And again, as capillary resistance and consequent high arterial tension cause cardiac hypertrophy by inducing over-work of the heart, we are warranted in concluding that the very opposite condition viz., greatly diminished tension, must lessen the labour of the central organ.

"Further evidence that there is a larger supply of blood at the surface of the body is afforded by an observation which I made on several persons, that after residing for some time at these altitudes, the face became decidedly fuller and redder. In some, the increased fullness or swelling was sufficient to produce a change in the expression.   That this was not owing to a gain in flesh was proved by its rapid disappearance on leaving the mountains.   Of course, this fuller and more florid countenance must be distinguished from the effects of sunburning, which, owing to the powerful rays of the sun

(a) *Glasgow Medical Journal,* 1885.

falling directly, and also reflected from the snow, is very common.

" Another proof of blood accumulating on the surface of the body from loss of atmospheric pressure, is the hæmorrhage which is apt to occur from the eyes or nose or ears during balloon ascents or in climbing lofty mountains."

Ozone.—This peculiar electrified condition of oxygen is formed in nature by atmospheric electricity, and is found in healthy localities—mountainous tracts, the sea-side, &c.—but scarcely exists in large towns, densely populated districts, or in the interior of houses, being used up and decomposed by animal and vegetable life. Since the discovery of ozone by Van Marum of Holland, 1785, and the researches into its properties by Schonbein of Bâle, in 1840, it has become to be considered a body of the highest importance in relation to health and the purity of the atmosphere, but no effects have hitherto been made to employ ozonised air in sufficient quantities to maintain a healthy and salubrious condition of atmosphere within large buildings. In a work on " Ozone and Antozone " Dr. Cornelius Fox states as follows : — " The salubrity of a town or city may be pretty accurately estimated by the effects of its air on ozonoscopes, as the feebleness and sluggishness of the reaction is a very good gauge of the amount of impurities which it contains. Ozone is a deodorising and purifying agent of the highest order, resolving and decomposing into innocuous forms. The oils of the cod's liver, the cocoa-nut, the sun-flower, *when ozonised*, have been found, by Drs. Theophilus and Symes Thompson, to be very useful in reducing the rapidity of the pulse, and exerting at the same time an invigorating influence on the heart's action in consumption. Last in order, but first in importance, ozone has been considered to be probably concerned in a work most gigantic in magnitude and of vital consequence. It has been thought to be influential in the modification of climate, to exercise a beneficial action on animal and vegetable life, and to be indispensible to the relief and cure of functional disorder and disease. It has been doubted whether life could continue to exist on this planet, according to the present constitution of terrestrial laws, if the formation of ozone should cease in nature.

It is, perhaps, too soon to say that ozone is in some way obnoxious to micro-organisms, nevertheless, the latter diminish with increase of ozone. In the elevated Alps, where there is constantly recurring atmospheric states of electrical high tension, large quantities of air become ozonised, and it is in these regions that micro-organisms have been found by Pasteur, Miquel, Tyndall, and others, to be greatly diminished in numbers and at some heights, as for instance on the Aletsch glacier, at an altitude of 2,300 mètres, Tyndall found that sterilised infusions were unaltered when the flasks were opened and closed again ; but, in a hay-loft, twenty-one flasks out of twenty-three showed living organisms, when manipulated in the same manner.

The researches of M. le Professeur Yung, of Geneva, and Edouard de Freudenreich of Bern, carry with them some valuable evidences of the purity of mountain air, in comparison with that of the plains. M. de Freudenreich has devoted much time to these important investigations. Many delicate experiments have been made by him at the Col de St. Théodule (3,322 mètres), Aletsch glacier, Niesen (2,366 mètres), Lake of Thun, &c.

The methods of procedure were similar to those originated by Miquel, of Paris, the director of the Montsouris Observatory. After demonstrating the entire absence of germs in the quantity of air examined (300 to 1,500 litres), at such altitudes as the Eiger (3,975 m.), the foot of the Eiger (2,100 m.), the Strahlegg and the Schilthorn, Freudenreich attributes the diminution in the number of microbes in the high regions, to the following causes :—

1. To the progressive disappearance of productive foci for bacteria, according to height, until the zone of eternal snow is reached, when the disappearance of these foci are absolute.

2. To the lessening density of the atmosphere, which becomes less able to hold microscopic particles in suspension; at the same time a diminution of dust from the same and other causes.

In an interesting *brochure* on "Living Organisms of the Air," he also mentions the recent researches of M. le Commandant Moreau, published in the *Semaine Médicale* of March 6th, 1884, by Dr. Miquel. These researches show that only five or six bacteria were found by Moreau, in sea-air.

Now, there can be little doubt from these and other observations that the relation of micro-organisms to ozone and sunlight is in an inverse ratio : for on the sea and in mountain air microbes diminish. Although temperature, a thin atmosphere, and absence of productive foci clearly affect the development of micro-organisms, the power exerted in this direction by an ozoniferous atmosphere is well worthy of careful investigation. It is well known that ozone has a remarkable effect on putrid blood in bringing it back from a decomposed state, into a condition free from offensive odour. In an experiment by Drs. Wood and Richardson where some blood had been kept for several years and had undergone the most offensive decomposition, a current of ozone soon restored this putrid dead fluid to so perfect a resemblance of fresh blood, that the mass coagulated and exuded serum. Dr. Fox states that ozone is thought by some to be absorbed by the blood corpuscles with great rapidity, oxygen being liberated. Kühne, on the contrary, is of opinion that the blood globules ozonise the oxygen with which they come into contact, without themselves undergoing any change.

E

CHAPTER III.

" Cold " as a Sensation and as Temperature.—Characteristics of Alpine Climate, and its Effects on the Vascularity of the Lungs and Skin, and on Nutrition.

THOSE who are unacquainted with high altitudes in winter may perhaps be inclined to judge the sensations at a high cold region from an English standard of cold, thinking possibly that twenty degrees of frost signifies twenty degrees of chilliness, and that any temperature below freezing point would be likely to cause discomfort to delicate persons. A brief explanation may tend to correct this view. The body can be deprived of its heat in four different ways :—

1. By conduction, or contact with colder substances, either solid, fluid, or gaseous.

2. By evaporation from the surface of the skin, and the mucous membrane of the respiratory tract.

3. By excretory matters leaving the body ; and

4. By radiation.

Now, although the hygrometric state of the atmosphere causes an additional quantity of moisture to be evaporated from the respiratory tract, at high levels, (*a*) this variation

---

(*a*) It must not be supposed that evaporation from the body depends entirely on the percentage of humidity in the air. The conservative and balancing agencies of physiological action amongst other things constringe or dilate the cutaneous capillaries in response to cold or heat, rest or exertion, atmospheric pressure, &c.

does not sensibly reduce the temperature of the body compared to the abstraction of heat by conduction; or in other words, contact with cold air in movement. This latter cause is the one which principally bears on the question of sensation, inasmuch as cause No. 3 is too insignificant to be felt, and No. 4 can be guarded against to a great extent by clothing.

The physical sensibility of cold is produced by the amount of heat rather suddenly abstracted from the body (which does not always depend on the temperature in contact with it). For example, if the hand be placed on fur it feels warm in comparison with iron at the same temperature. The former being a bad conductor—owing to the *motionless* air in its interstices—does not abstract much heat from the hand; the metal being a good conductor of heat, appears intensely cold to the touch.

If, therefore, cold—motionless—dry air surrounds the body, heat is not abstracted nearly so readily as it would be by somewhat warmer air in movement. It must be remembered that the sensation of cold cannot be accurately gauged by reference to the thermometer. Two other conditions are intimately connected with temperature in causing impressions of cold or heat—viz., wind and moisture, for it is these that cool the body by conduction. If their temperature is lower than ours, they appear colder than they really are, because from their conductivity heat quickly passes away from us.

In the high valleys of the Alps, although the thermometer may register some 15° or 20° of frost, this low temperature is by no means disagreeable in calm air, as the intense solar heat enables many persons to sit in the open and bask in the sun during the depth of winter without feeling the slightest sensation of chilliness. Even excessive tanning and reddening of the skin takes place with almost everyone who takes plenty of outdoor exercise: ladies, who are generally well protected by sunshades or umbrellas, do not

E 2

escape a healthful aspect. This is mostly owing to the reflection of light from the snow, which coming in upward and parallel directions, cannot be well screened from the face. The habit and necessity of wearing smoked-glass spectacles also enables persons to face the glare, and thereby receive a much larger proportion of light than in England. (a) According to Dr. Cornelius Fox, ozone also causes a healthy colouration of exposed parts of the body. The speedy tanning that one undergoes when crossing a mountain pass, or driving in an open sleigh, even in dull weather, supports this view. As more ozone is brought into contact with the skin by movement through air when driving, or on the passes where there is generally a breeze present, so a greater colouration ensues : whereas one may be exposed to the influence of wind in large cities (where there is an absence or great diminution of ozone) without wearing a rosy or tanned appearance.

The characteristics of Alpine climate are—

1. Dryness of the air (b) and its comparative freedom from

(a) It has been frequently noticed that dark-complexioned individuals become sunburnt more readily than "blondes." This depends principally on the sensitiveness of the retina and the colour of the eyes. For instance, "fair" people cannot face the light with such ease and comfort as those who have plenty of pigment in their ires ; for the pigment absorbing the rays of light, protects the retina, and even enables some very "dark" eyes to gaze on the sun itself. On the other hand, a person with a grey or pale iris averts and screens the eyes from the sunlight as much as possible, and in this way escapes the effect of the rays on the face.

(b) The drying and preserving of meat hung in the air has been alluded to as an illustration of the dryness of these climates. This takes place in an analogous manner to the drying of turtle in the sun in the West Indies, viz., evaporation of moisture before putrefaction takes place. In the Swiss mountains the vitality of germs being rendered inert by a low temperature, thick masses of meat can be gradually dried.

micro-organisms, mechanical irritants, and noxious gases.

2. Profusion of sunlight.
3. Diminished barometric pressure.
4. Ozoniferous atmosphere.
5. Low temperature.

The results on pulmonary complaints may be stated thus :—

1. Increased chemical action in the blood and tissues, from sunlight and reduced atmospheric pressure.
2. By breathing aseptic air free from dust, irritation, or perhaps recurrence of infection by microbes in the respiratory tract, is greatly lessened.
3. Vaporisation of morbid secretions in the lungs, promoted by reduced barometric pressure and dryness of the atmosphere.
4. Increased quantity of blood circulating in the lungs—caused by the low temperature—the freedom of the circulation being aided by extended chest movements.
5. Increased activity in the pulmonary lymphatics (depending on circulation and expansion) and a general improvement in nutrition and glandular secretion ; also an exhilarating effect on the nervous system.

With regard to the increased quantity of blood circulating in the lungs (presumably influencing the nutrition of those organs), it may be contended that this is not a desirable sequence. Perhaps it is not in hæmorrhagic phthisis ; but in some other forms, especially early tubercular deposits, it would not seem to be disadvantageous. What would lead one to suppose this, is the rare occurrence of tubercular phthisis in persons affected with mitral disease. Even when hæmoptysis takes place, and when some of the blood presumably gravitates into the air cells, tubercular disease rarely

follows; whilst, on the other hand, phthisis is not an uncommon consequence after hæmoptysis from other causes.
This would appear to indicate that a general hyperæmic condition of the lungs impedes the deposition of tubercle, and
restrains phthisical processes.

Conversely, where the quantity of blood circulating in
the lungs is diminished, as in hot climates, phthisis is
frequently seen to run a very rapid course.

On the other hand, the emphysematous signs presented by
patients who may be said to be cured, after a prolonged
residence at high stations, may seem to contra-indicate any
theory based on this assumption. It must be conceded,
however, that with the expansion of the chest obtained, it is
doubtful if emphysema occasions such compression of the
pulmonary capillaries as to decrease the *whole* volume of
blood circulating in the lungs.

The liability, also, of the natives of these high valleys to
pneumonia, whilst exempt from phthisis, would seem to point
to some alteration in the vascular condition of the organs
affected. What result any variation in the vascularity of the
lungs would have on the bacillus tuberculosis is rather prema-
ture to surmise. No bacilli have, however, up to the present
time been discovered in the blood of tubercular subjects. It
appears, therefore, that either they do not enter the vessels in
the form of a bacillus, or if entering, are changed in character
or destroyed. That the state of the blood, chemical or pathological condition, or functional activity of the tissues must be
agencies governing the suitability of the soil for the reception
of any micro-organism, is supported by the fact that infection
is very rare, although there are numberless cases in which
bacilli have undoubtedly been inhaled.

There is, however, no decided criterion for the determination of the question whether a larger quantity of blood in
the lungs, such as may be diffused through these organs in a
high cold climate, with accelerated respiratory and cardiac

movements, is or is not, disposed to bring about pulmonary hæmorrhage.

It is, however, a fact, supported by many observations, that hæmorrhage from the lungs does not occur so often at high elevations as to be a contra-indication against sending cases of hæmorrhagic phthisis to mountain resorts. Dr. Charles Denison, of Denver, declares that hæmorrhagic first stage cases constitute a most favourable class of patients for high altitude treatment, and his observations, made at heights much higher than Davos Platz, are in accordance with my own experiences and those of others, for the Swiss altitudes.

It has been shown that the hygrometric and barometric states of the atmosphere modify the process of evaporation from the lungs and skin. The evaporation of morbid secretions in the lungs was pointed out by me in 1881 (a) as being one of the circumstances which probably has an important bearing on phthisis. The process of evaporation in dry climates, acting on ulcers, cavities, or suppurating surfaces, if not analogous to the dry treatment of wounds, brings about a less moist and watery condition of the secretions from diseased bronchial tubes or cavities of the lung; virtually imitating expectoration without the patient undergoing the effort of coughing.

The effects on the body of sunlight and reduced pressure are those facilitating chemical action in the blood and tissues, whilst the cold air necessitates the requirement of a larger absorption of oxygen and assimilation of hydro-carbons to maintain the heat of the body. It may be conjectured that this contributes to the sudden and considerable push given to nutrition on arrival at a cold high altitude, when the appetite is, in most cases, at once improved in a remarkable way, and animal food that could hardly be thought of

(a) "Davos Platz, and the Effects of Altitude on Phthisis."

previously without disgust is eaten promptly. This push to nutrition is a reliable feature in the first evidences of progression, and in conjunction with an early reddening or tanning of the skin, assures a certain measure of hopefulness in the case.

The exhilarating feeling produced by the *consciousness* of moving about amid snow and ice, without taking cold or feeling pinched, is not to be despised as contributing towards cure. The contrast of this with the life in England during winter, where every change of weather has to be guarded against, is so marked that the hope of recovery presents itself, and despondency is banished.

There is not much to be said on the subject of electricity in relation to climate, as very little is known ; my notes on atmospheric electricity are not yet sufficiently complete to justify anything but the statement that the dryness of the air in the Swiss Alps favours electrical commotions. The slightest friction of clothes, walking across a carpeted room, the impact of air on the body, &c., will give evidence of changes being evoked in the electrical condition of the surface of the body. This can be tested clearly by a condensing electroscope. Electric tension, too, is more marked in dry mountainous regions, giving origin to the formation of large quantities of ozone.

There is every reason to suppose that under the many favourable circumstances presented by these climates the treatment of suitable cases of anæmia and its allies, scrofula, consumption, affections of the chest, and some cachectic states of the system, can be undertaken with greater confidence, and those measures which have of late years prolonged many valuable lives, are certainly more likely to be efficient and successful, when supported by the curative effects of mountain air ; whilst for the somewhat minor maladies, such as debility, either from physical causes, or from mental fatigue and worry, malaria, &c., some forms of dyspepsia,

chronic discharges or suppurations (that do not incapacitate
the patient from taking gentle exercise), and during convalescence from many acute diseases, the renovating power of
these climates are, in suitable cases, doubtless superior in
rapidity of effect, to warmer and lower latitudes.

It is not implied that climatic conditions, grateful at all
times, are not found in certain localities, where the range of
the thermometer is generally from 50° to 70°, or thereabouts.
This equality of temperature enables patients to be constantly in the open air, and if not out of the house, to be
able to avail themselves of a system of "hyperventilation"
day and night; but, on the other hand, it is now well known
that other places can be found as favourable, if not more so,
for these complaints in cold climates, although the range of
the temperature is not so limited; whilst the languor and
depression of strength felt in warm climates is altogether
escaped.

Taken as a whole, mountain climates are exciting and
stimulating, and the mental effect of the sunlight on many
patients who have been habitually spending much of their
time indoors at home, is most beneficial. The value of
exposure to fresh air in many diseases is becoming more
widely known and appreciated, and its benefit to phthisical
cases, provided they escape attacks of catarrh and bronchitis,
cannot be doubted, whilst the tonicity of mountain air
in anæmia, debility, dyspepsia, some uterine complaints,
and convalescence from acute diseases, are too well known by
Anglo-Indians to require repetition.

The value of cold high climates in pulmonary complaints
is, in a great measure, the consequence of a gentle training of
the respiratory functions, the lungs becoming expanded
without calling for the usual stimulus of muscular effort,
which in many conditions of disease would be injurious or
perhaps dangerous from the consequent force given to the
pulmonary circulation.

CHAPTER IV.

Including the Drawbacks of the Alpine Winter Stations.

AFTER leaving Coire (a) for the ascent to Churwalden (4,976 ft.) and Parpan (4,948 ft.) the beauties of the mountain journey begin. The rise in the altitude gives a sensation of lightness and elasticity to the frame. All the surroundings cheer the spirits—the tinkling music of the cow-bells, the autumnal changes in foliage, golden larches and aspens contrasting with the dark green of the firs, topped here and there by brilliant-sunbeams, undulating into gorges and hollows of impenetrable shades—whilst in the distance are occasionally seen the higher peaks crowned with dazzling snow. After a railway journey, perhaps from foggy London, these scenes alone stimulate and recruit the strength. It is well to be cautious at this stage, that the powers are not overtaxed by fatigue; for this sudden vigour is not yet permanent, but may permit a delicate person unconsciously to out-run the supply of force. It is, in some instances, desirable that the journey from London, for example, should be broken at various points, so that the patient shall not reach his destination in a wearied and jaded state.

(a) Davos Platz and Davos Dörfli can be reached from Landquart, the station before Chûr (*vide* page 23).

COIRE.

The transformation scene after the winter snow has fallen intensifies all the virtues of mountain air. With the sudden covering of the earth by a non-conducting material an entirely new condition of the climate is established. Fresh breezes have died away, packs of cumulus cloud have been condensed into snow, the moisture in the atmosphere is of the smallest quantity, whilst the sun shines with the lustre and power of the South, and the sky presents an almost unclouded surface of azure blue.

One of the drawbacks to some of these health-resorts, as with many others elsewhere, is defective sanitation. As long as overcrowding does not take place, the cold and snow will, to a great extent, mitigate the evils of bad drainage and impure emanations; but when hotels get filled with a large number of visitors, many of whom may be in delicate health, the air within will not (except by artificial ventilation and good house drainage) be free from the usual indoor impurities, as sewer-gas, kitchen and basement air, exhalations from the lungs and skin of organic matter, scales of epithelium, fibres of cotton, wool, wood, &c., the products of combustion from gas, lamps, and candles; bacteria and fungi; and, what is, perhaps, more important still in the case of delicate lungs, bacilli floating about in the air. Patients must, therefore exercise their own discretion in opening windows and airing their rooms as much as possible. During fine weather a fair quantity of fresh air can always be·permitted to enter the bed-rooms, according to the desire of the individual; but in other parts of the houses the means of admitting pure air and providing for the escape of foul are not always very perfect. Although Tyndall has shown that expired air is "optically pure," and the researches of MM. Strauss and Dubreuille by bacteriological methods, have demonstrated that expired air is almost completely deprived of microbes, it has been frequently proved and is well known that poisonous products are exhaled from the lungs, and that even if organic matter in the air leaving the bronchial passages is not already putrefied, it rapidly undergoes noxious alterations.

Where stoves are in use, attention should be given to them to see that servants do not entirely close the valve which regulates the flue. Occasionally this is done, when the fuel is consumed and the remains are glowing red, with the intention of economising heat, which it effectually does, by preventing air passing through and cooling the stove; but the most poisonous of all gases, viz., carbonic oxide, is slowly evolved into the room from the dying embers and quickly gives rise to headache, which is not rapidly shaken off even by fresh air. (a)

The purity of air in the interior of dwellings demands every scrutiny when we take into account the number of hours spent indoors. On the shortest day of mid-winter the sun remains but five or six hours in the valleys, therefore delicate persons must be content to pass 19 or 20 hours in the house. It is obvious that some of the benefit derived from open air exposure will be neutralised unless constant currents of fresh are supplied for respiration.

It is essential, too, that some special employment should be undertaken when persons remain several months abroad; languages can be studied, or the mind occupied with work of some sort. By this means time does not drag heavily, *ennui* is not experienced, and restoration to health is not retarded be dejection of spirits.

The travelling also has to be considered. Those who are in pretty fair health can make the journey in three days to the Engadine, but a longer time is recommended for those whose health does not admit of prolonged confinement in railway carriages. The journey may be broken at Paris or

(a) Carbonic oxide replaces oxygen in the blood, and cannot be again replaced by oxygen, but has to be slowly converted into carbonic acid, before elimination.

Brussels, Bâle, and Chûr, Como, Chiavenna, or Promontogno, according to the route chosen by the traveller.

On first arrival it occasionally happens with some people that they are unable to obtain a good night's rest for the first night or so. This depends on the sudden rise which they have made to high regions. The difficulty soon passes off; but in the case of sensitive persons who sleep badly, it may be met by remaining a few days at Chûr (1,936 ft.). The stay there will frequently help in training the system to the new conditions. It must not be forgotten that the altered form of bed may, in many instances, prevent sound and refreshing sleep. The wedge-shaped hair bolster can sometimes be removed with great advantage to the sleeper, and extra clothing be placed over the shoulders and upper part of the body which is usually not covered by the cider down quilt. In addition to equalising the bed temperature in this way, the head should not be raised at too great an angle from the chest and shoulders, or compression of the vessels of the neck will ensue, and so impede the ready return of blood from the brain, keeping this organ active and full, when it should be in the anæmic state which usually accompanies sleep.

Discomfort is sometimes caused by chilblains during the winter time. To prevent these, walking exercise should be taken immediately after breakfast, and the endeavour be made to keep the feet warm throughout the day by this means. The parquet floors, although healthy and advantageous in many respects, are, in a great measure, the cause of cold feet and the source of chilblains, therefore all the corridors and bed-rooms in hotels should be partially carpeted in these climates during winter time.

What has been advanced as an objection to Alpine stations is the use of the German stove as a means of warming the interior of houses and hotels. The fault, however, does not lie so much with the ponderous German *Ofen* as with improper management in neglecting to provide sufficient ventilation for the rooms and moisture for the air. An improvement on these stoves are the steam reservoirs in use in some houses, the amount of heat being under the control of the occupant of the room in which they are placed. But this plan also raises the temperature without raising the "dew-point," or, in popular language, "dries the air," creating a necessity for a supply of watery vapour to render the atmosphere fit for healthy and agreeable respiration ; for, although dryness of the climate is one of the main features of the Swiss Alps, a limit can be reached beyond which a dessicating effect may be produced, with laryngeal irritation.

The introduction to the bright and calm winters of these regions is sometimes a little unsettled, but not dangerously so. Falls of snow occur and are apt to thaw again ; the roads, however, soon dry, by reason of the sloping land and rapid evaporation. Acclimatisation has frequently been held up as a pre-requisite for visitors, but, as the winter is generally bright, calm, with dry cold and plenty of sun, the risk of arrival, even after the commencement of winter, is in no wise greater than during the autumn. To become acclimatised for one of the best seasons of the year, and one which is regarded as a "cure" period, requires no inurement —many persons even arriving in mid-winter and crossing a mountain pass without inconvenience or harm resulting. The selection of an Alpine station should certainly be influenced by the opinion of physicians at home, even where the intention is entertained of returning for a second or third winter. Perhaps this may seem to be written in the interests of the medical profession, but those who have spent several seasons at these places will have observed that some cases return again who would do better in a different climate. An over-estimation of their own recuperative powers, or an enthusiastic belief in the potency of the charming Alpine climate to sweep away *all* maladies, is possibly the cause of an injudicious decision.

Children over three years of age do extremely well in the Alps, make healthy blood and muscle, gain flesh, and expand their chests. Under this age, it is doubtful if sufficient exercise can be taken by a child to ward off the cold, without being swathed to such an extent in furs and flannels as would impede the free motions of the limbs and thorax, so essential to development in childhood.

It is also a moot question whether some invalids should not quit the Alps immediately the end of the winter seems approaching, as the changeable weather, with wet roads, winds, &c., is likely to upset many of the benefits gained during the dry season. This is one of the great drawbacks of all European health resorts—namely, that a change in spring time is unavoidable.

It has been proved now by experience that all relaxing climates, moist warm valleys, the neighbourhood of lakes with the evaporation that constantly takes place from a large expanse of water, do not act favourably on delicate persons who have been living for several months in the dry air of the Alpine heights, A warm moist air or a rapid descent to a low level must be avoided as much as possible, and the return to England should be made as late as circumstances will permit. If it is as late as May (depending a good deal on the weather) so much the better ; or if a prolonged stay for the invalid has been ordered, he can again return to the heights after the snow-melting has entirely gone.

Wiesen (4,771 feet) may be said to be one of the best halting places for those returning from the Engadine. Owing to the sloping and sunny aspect of the Wiesener Alp, the snow disappears very early there, even some weeks in advance of many places at a lower altitude. Often whilst the Upper Engadine is still covered with one or two feet of packed crystallised snow, Wiesen is quite clear, the post running on wheels instead of sleigh, and the country around putting on an aspect of spring. Snow-falls nevertheless occur sometimes in April, therefore winter garments must not be laid aside at the sight of crocuses, flowering heath, and anemones.

Thusis (2,448 feet), has attracted many persons from Davos, it is a pretty place, with agreeable walks and fairly comfortable hotels. Ragatz (1,709 feet), Mels (1,637 feet) are reached easily, and a residence at the former has become rather fashionable after the cold season. Promontogno (2,700 feet) is very convenient for the Engadine if a return is to be made by Italy, its position and climate go a great way in recommending it as a halting-place: measures are being taken by the proprietor of the hotel there to warm the rooms and corridors, as it is well-known that much depends on the comfort and equable temperature within doors during spring-time. Coire (1,900 feet) according to Dr. Killias, shows a dry climate in the spring, and many who have descended from Wiesen there, have done extremely well, and have found an absence of that relaxing feeling so often experienced on reaching a level of about 2,000 feet. It is not recommended that frequent visits be made within the town itself. There are several picturesque walks a short distance from the principal hotels.

		Station.	Barometric Height.	Mean Temperature.	Mean Cloudiness	Rainfall in Millemètres.
					per cent.	
1872	March,	Thusis.........	746ᵐ	9.11	56	214.9
to	April,	Chûr	603ᵐ	9.27	63	149.7
1875	and May	Ragatz.........	541ᵐ	9.22	44	335.9

A month might be spent at any of these, previous to returning home, but it is well to be impressed with the necessity of the great personal care needed during this migratory period, especially with regard to clothing. Thick flannels, socks, &c., should not be dispensed with hastily, and it would be well to guard againt the change from the calm regions by adopting outer garments of close texture, impervious to the chilling effects of wind.

Any remarks on the climate of Switzerland would be incomplete without reference being made to the föhn wind.

Many nonsensical theories have been advanced as explanations of this wind—one, that the dry air comes from the Sahara Desert, is revived from time to time. What really takes place in Grisons is shortly this :—A south-west wind strikes the southern declivity of the Alps, and ascends into the higher regions of the atmosphere pressing the air down into the narrow valleys. In proportion to the narrowness of the valley, so is the insalubrity of this unwholesome wind, which is in many instances quite a local phenomenon, depending on a current of air of limited extent or force. Altitude, temperature, season of the year, configuration of the valley, and altitude of the main current of air above, modify to a great extent the noxious qualities of this wind, and what would be a depressing enervating föhn in the bed of a narrow shut-in valley, would, on a higher and more open elevation, be a cooler wind, and more free from disturbing influences as regards the nervous system.

The föhn then is a moist southerly aërial current. Its temperature in Switzerland becomes elevated from various causes, the chief of which is atmospheric pressure. By this accession of heat the capability of the föhn for holding more moisture in suspension is at once augmented. The "absolute" humidity may remain the same, but by a rise in temperature the percentage of relative humidity, i.e., humidity in relation to the point of saturation), is considerably reduced, for warm air can contain more watery vapour; therefore, the föhn seizes with avidity on any

F

moisture which is present. The extreme dryness is again advanced by the stoves in use, further raising the temperature to a dessicating point. In this way many bad symptoms are aggravated and vaporisation of organic matter takes place, which is doubtless re-inhaled by the occupants of a room badly ventilated or unprovided with sufficient water near the stoves to bedew the air.

The nervous system of most persons becomes remarkably depressed by this wind; the inclination to undergo exertion is diminished; sleep and digestion are disturbed; the animals, too, seem to suffer—effects which closely resemble those caused by the "Vent d'Espagne," in the Pyrenees (Hermann Weber). Indisposition is usually attributed to this wind at Davos, and in several instances with justice.

After one or two days, sometimes longer, the second phase of the föhn is developed. The thermometer falls and with the cooling process the air is gradually brought below the dew-point, and what a few days before was an exceedingly dry wind, becomes converted into an atmosphere saturated with moisture, which falls as snow or rain, according to the time of the year, and temperature.

The days of invasion by the föhn wind are about equal at Davos and Wiesen. On the Maloja plateau, 1,000 feet higher, this wind is rarely felt, and its noxious qualities are greatly modified by the higher altitude, extended width of the valley, and the proximity of immense glaciers to the S. and S.W. At St. Moritz Kulm, which has still a higher elevation, there is the same freedom from the pronounced nervous derangements caused by this wind.

The teeth are apt to cause some trouble in the Alps, if any disposition to caries be present. A good dental surgeon should be consulted previous to leaving home, and the habit be acquired of breathing through the nose as much as possible, even when taking exercise. The rapid changes in the temperature of the teeth occurring to persons who breathe with the mouth, give rise to a variety of derangements, especially in cold climates. Among the natives of Switzerland the noticeable deterioration of the teeth may probably be due to the consumption of the rough wines of the country, which doubtless contain much acetic acid, caries being very marked in the lake districts.

There is too, a condition of pyrexia sometimes met with at high elevations, which may be termed mountain feverishness. It has been well described by Dr. B. P. Anderson, of Colorado Springs, but as far as my own observations go, I have never seen the type of so severe a character as mentioned by him and others.

A sudden invasion of high temperature running up to 102° and 105° Fahr. is sometimes experienced, with hot skin and irritable heart. The pulse varies from 80 to 112 and although affected by movement, as are also the respirations, they do not follow the temperature. With this there is gastric irritabilty, but no vomiting nor extreme loss of appetite, the tongue is slightly furred, moist and of a natural size. A sense of fulness in the head is more marked than actual headache, there are slight creeping chills, tendency to constipation and thick urine. No deafness, tympanitis, abdominal pain, nor rash. The duration of this condition is about one week, with the tendency to relapse. Towards the end of this time, perspiration occurs, the appetite falls off, and some weakness is felt, but the strength and pulse are kept up out of all relation to the temperature.

The chief points which distinguish this pyrexia from simple continued fever, are :—

1st. High temperature out of all proportion to constitutional disturbance.

2nd. Irritability of the stomach and indistinct craving for food, possibly due to increased secretion of gastric juices.

F 2

3rd. Suddenness of attack, remittent character and tendency to relapse. History of exposure to sun-heat and over-exertion, especially during snow-evaporation.

It is an erroneous notion to suppose that there is absolute immunity from consumption at heights of 5,000 feet, where there is a predisposition to the disease. I know of cases which have originated *de novo* at Davos, but amongst the natives, as far as it is possible to ascertain, immunity has really existed, up to the present.

In conclusion, it may not be superfluous to remind those interested that the high cold regions are not yet proved to be the best for *all* cases. The varieties of chest and other affec- tions and the individual differences of affected persons denote that one general climatic panacea is quite inadmissible, and what may be a suitable climate for one, may prove of very little value to another. Whilst some misconception prevails on these points, cold mountain air will not receive the appreciation it merits. It would be inconsistent to contend that any particular health resort had not its drawbacks. It is misleading to make light of these, and most essential in the recommendation of foreign health-resorts that their disadvantages be clearly stated, as well as the probable benefits likely to accrue to persons visiting these places. Much disappointment is thereby averted, and many, who otherwise would expect impossibilities, do not become discontented.

The accusation that mountain climate is occasionally re- commended or adopted without sufficient judgment and discrimination, and that the snow-lands of Switzerland are unsuitable for a large number of phthisical cases, is not unfounded. It will be seen further on that many circum- stances, symptoms, and habits of life, have to be taken into account before a winter residence in the mountains can be regarded as the best means towards recovery. The advantage, then, over southern health resorts or the ocean, are more accurately appraised by those medical men with personal experiences of each or any climate they advise to their patients.

The best description of the ocean voyage is to be found in Dr. J. A. Lindsay's book on the "Climatic Treatment of Consumption." What he says carries all the more weight with it, as it is the outcome of actual experience of the climates he writes about.

A word of caution is needed to those sick persons who, having got tired of medical treatment at home, decide to drop "doctoring" and go in for climate and self-medication. If these people select the winter of the Alps for this experiment they choose one of the most potent climates that can be fixed upon, and where guidance by an expert, if not indispensable, is of the greater value to anyone in delicate health.

Something may be said also on the length of time which it is desirable to spend at high Swiss levels. Two years continuous residence without a break is to be deprecated, and it is questionable if all the best results of these climates are not exhausted after four winters, even for the major number of pulmonary cases. It will be found that a change to a new climate after this time will be of more value. In many cases other than lung troubles a longer residence leads to loss of flesh, dyspepsia, nervousness, or inability to sleep well, and an intolerance of stimulants, such as wine and tea; and there are periods when a change to a lower level is clearly indicated, although seldom recognised.

In spite of much misunderstanding and even opposition, high altitude treatment is gaining ground, and will, no doubt, take its right place in climatic therapeutics. Time is needed for the subject to be well threshed out by advocates and opponents alike.

CHAPTER V.

Winter Clothing.

THE selection of suitable clothing for an Alpine climate will contribute greatly to health as well as to comfort.

The tendency is generally to imagine that thick heavy cloth garments and solid jackets, which button up over the chest, are essential to keep out the cold. These have all to be cast aside on personal acquaintance with the climate, and hare-skin, flannel, or wash-leather "chest-protectors" must all be banished to the rag-bag as worse than useless.

The first advice tendered to ladies who seek restoration to health from lung disease is:—Abandon corsets absolutely, and wear the loosest *ceinture* on the waist. This injunction cannot be emphasised too strongly. One prominent sign of what may be termed "cure" of phthisis, at high altitudes, is the almost invariable expansion of the whole or portions of the chest walls. To impede such a beneficial result is to limit the advantages of breathing mountain air. No one could wittingly be guilty of such an irrational procedure as to curtail the actual physical improvement they travel many miles to secure. Yet usually, whilst the upper regions of the chest are allowed free play, the most mobile section, the diaphragm, is fettered and confined in its movements by an unyielding investment pressing on the abdominal viscera. In this way an opposite conformation is maintained against what physiology and common-sense would indicate ; for in many instances where damage exists in the superior thoracic regions, limitation of movement is desirable in *these* situations, rather than in the lower and sounder parts of the lung.

The substitute for corsets should be a thick flannel waistcoat or jersey. Woollen garments should also be worn next the skin over the trunk and extremities. Chamois leather may be substituted when over-sensitiveness of the skin renders the use of wool unbearable ; generally, however, a fine texture can be procured which is not too irritating. A good stock should be taken, as frequent changes are beneficial.

The coverings for the feet and legs require attention. In all cases thick socks or stockings are indispensable, and pants should be made long enough for socks to be drawn over them. By enveloping the whole body in this material the patient is spared the necessity of loading the exterior with thick weighty dresses or heavy overcoats, which fatigue the wearer and confine the act of respiration. Those who are liable to suffer from cold feet, or are susceptible to chilblains, will find thick worsted socks and cloth gaiters a necessity, or thickly lined cloth boots, to come well up the leg. As a prophylactic against chilblains, bathing the feet in salt and water, avoiding tight boots, wearing digitated socks, and changing them whenever the feet are damp, keeping the arms warm and the hands dry, will carry many through the winter of a cold climate, who might otherwise suffer much inconvenience and pain from these minor ailments.

Ordinary stout skating boots, with broad toes, low heels, and plenty of room inside them, which may be filled up with a cork sock, are suitable both for ladies and gentlemen ; they should be well greased every day and always changed immediately after a long walk. Goloshes with cloth tops are useful for short journeys or sitting out, but cannot be recommended for long walks on account of confining the perspiration. All clothes should be light in colour, as near grey as the taste will allow. A light colour does not absorb nor radiate heat so much as black; therefore a wearer of grey will be comfortably free from the heat of solar radiation, and warmer in the shade. Waistcoats should be lined in the back with flannel, as with modern garments all the protection is in front of the body. (a) If men endeavoured to dispense with the use of braces greater play would be given to the muscles of respiration. Hats may be of straw for the sunny weather, but light-coloured thin felt answers very well. Furs should not be worn when taking exercise. The ladies' fur tippet can be discarded as superfluous at any time, from the fact of its covering the shoulders only, causing that part of the body to become over-heated, and therefore liable to chill. A short list is appended of article found to be serviceable for these climates :—

One fur or railway rug. One warm ulster, with hood or fur-lined coat for travelling and sleighing. One thin overcoat, a Shetland shawl. Two or three suits of clothes. Thick woollen vests and flannel shirts. Flannel night-shirts or suits of pyjamas. Worsted socks. Pants of wool, not merino. Cloth or fur gloves with gauntlets are useful for sleighing, and cloth or dog-skin driving gloves for tobogganing. Cork socks. A pair of dark neutral tint spectacles. A few rough bath towels and a flesh glove. Two pairs of stout boots (one pair smooth for skating, the other with a few spikes in the soles). A pair of leggings are required for " coasting or tobogganing," unless long boots are taken, long spatts of thin box-cloth are the best, to be worn with pantaloons. Ladies can wear leggings or long spatts over thick stockings. Dresses for out-door exercise should be cut short. Divided skirts are excellent. One or two pairs of

---

(a) With double-breasted apparel a man may have from ten to thirteen thicknesses of woollen clothing in front of him, whilst his back is covered with three layers of wool only.

shoes, or slippers with heels can be worn indoors with short spatts over them. Bootlaces, dubbin, and skates may be added, and for ladies a sunshade and goloshes with cloth tops A fur foot-warmer or large fur-lined boots are also very serviceable for sleighing.

Diet in the Swiss Alps.—The food almost everywhere in Swiss hotels, consists largely of azotised or nitrogenous matter. Many who have visited Switzerland will recollect that much lean meat is eaten, whilst the proportion of animal fats is very small.

It is essential to the health and well-being of individuals that a proper proportion of nitrogenous and non-nitrogenous food should be consumed. The latter consists chiefly of starch, fats, sugar, saline substances, and water, which in the ordinary way with meats form the mixed dietary most suitable for man ; as the structure of his teeth and past experience indicate. But whilst an almost exact estimate can be formed of the elements necessary for the system to ingest, it must not be forgotten that the quantity and kind of food taken will depend very much on the condition or idiosyncrasy of the consumer. As a general rule, few delicate persons would be capable of effectually digesting the quantity of bread, pastry, milk, and root vegetables, &c., which would be requisite to constitute with the meat eaten, a fair combination of nitrogenous and non-nitrogenous matter, and as there is a marked absence of palatable fat at these places, a little consideration of the means for balancing the needful combination, will conduce to the recovery of health and avert many ill-effects arising from perverted nutrition—as dyspepsia, troubled sleep, a loaded tongue, vitiated secretions, &c.

"Many people," Dr. Pavy remarks, "seem to look upon meat almost as though it formed the only food that really nourished and supplied what is wanted for work. The physician is constantly coming across an expression of this view."

The greatest importance must be attached to the use of fats during winter in the Alps; for it is well-known that the inhabitants of Siberia, Greenland, &c., and all cold countries, eat enormous quantities of these heat-producing materials, without which they would be unable to resist the intense cold of the frigid climate. Sir John Ross observes—"It would be very desirable, indeed, if the men could acquire the taste for Greenland food, since all experience has shown that the large use of oil and fat meats is the true secret of life in these frozen countries."

Dr. Cheadle also lays much stress on the value of fats in cold climates. "One effect of the cold was to give a most ravenous appetite for fat. It is the most valuable part of food in winter, and horses and dogs will not stand work in the cold unless fat." (a)

Besides forming the chief articles of diet which are required for a calorifacient or heat-producing agent, they may almost certainly supply a pabulum for the oxidising process of fever in phthisis, to act upon, thereby not only restricting the waste of tissue, but perhaps in some way diverting its occurrence, as is witnessed when cod-liver oil is taken. With an increase in animal food, which can be readily eaten at high cold levels, the appetite being sharpened by exercise and low temperature, the necessity for fats is by no means abated; on the contrary, a physiological demand is created for additional food capable of undergoing the process of oxidation, which cannot be wholly supplied by the lean meat or by starchy and saccharine bodies.

"It appears from the experiments of Pettenkofer and Voit that increasing the proportion of nitrogenous matter in the food determines an increased absorption of oxygen by the lungs. Nitrogenous matter it is which starts the changes

(a) "The North West Passage by Land." Viscount Milton and Dr. Cheadle.

occurring in the system, and the suggestion presents itself that upon the amount of nitrogenous matter may, to some extent, depend the application of oxygen to the oxidation of fatty matter. Under this view the success of Mr. Banting's system may be due, not exclusively to the restriction of the principles that tend to produce fat, but in part, also, to an increased oxidising action promoted by the large amount of nitrogenous matter consumed." (Pavy.)

Let us draw attention to the substances at these health resorts that furnish the calorifacient group of alimentary principles. Disregarding the fat produced by a complicated metamorphosis of the carbo-hydrates and a small part of the nitrogenous food ingested (a) the main articles of diet from which fat is derived, would be butter and milk. About 1 oz. of the former at breakfast, and 2 pints of milk during the twenty-four hours together furnish at the most 2·6 ozs. of fat, allowing from ½ oz. to 1 oz. for that contained in the lean meat with gravies, &c., 3 to 3½ ozs., are obtained. It is doubtful if this is sufficient with exercise and low temperature, nor is it to be recommended that those with poor appetites should drink more milk. Persons should eat plentifully of butter at breakfast, especially as that meal is not a substantial one. Fat bacon can be added with advantage.

It would also be very advantageous to continue taking cod-liver oil if it had been found to agree in England, for reasons that are plainly apparent. Should the stomach have been unable to digest it hitherto, trial should again be made in a cold climate, as it might be then more easily assimilated, its oxidation being assisted by the increased proportion of nitrogenous material ingested. A good time for taking it is

(a) It is questionable, if with the low temperature in the Alps, phthisical persons should be physiologically compelled to maintain much of the body-heat by the carbo-hydrates, as the changes in their principles previous to their becoming calorifacient take place in the liver ; and the function of that organ is often impaired in phthisis.

immediately after lunch or dinner in a glass of marsala, or in milk half an hour after meal, commencing with one teaspoonful, and gradually increasing the dose.

To further demonstrate the virtue in cod-liver oil, fat, and butter, the following table taken from Frankland, will give a clear idea of their force-producing value :—

Name of Food.	Per cent. of Water present.	In units of heat.	When burnt in oxygen.	When oxidised in the body.
			Force-producing value.	
			In kilogrammétres of Force.	
Cod-liver oil...	—	9107	3857	3857
Beef-fat	—	9069	3841	3841
Butter	—	7264	3077	3077
Cocoa-nibs	—	6873	2911	2902
Cheese (Cheshire)	24	4647	1969	1846
Isinglass	—	4520	1914	1550
Bread-crust	—	4459	1888	—
Oatmeal	—	4004	1696	1665
Flour	—	3936	1669	1627
Pea-meal	—	3936	1665	1598
Arrowroot	—	3912	1657	1655
Ground rice	—	3814	1615	1591
Yolk of egg	47·0	3423	1449	1400
Lump sugar	—	3348	1418	1418
Grape sugar (commercial) ...	—	3277	1388	1388
Hard-boiled egg	62·3	2383	1009	966
Bread-crumb	44·0	2231	945	910
Ham, lean (boiled)	54·4	1980	839	711
Mackerel	70·5	1789	758	683
Beef (lean)	70·5	1567	664	604
Veal (lean)	70·9	1314	556	496
Guinness's stout	88·4	1076	455	455
Potatoes	73·0	1013	429	422
Whiting	80·0	904	383	335
Bass's ale (alcohol reckoned) ...	88·4	775	328	328
White of egg	86·3	671	284	244
Milk	87·0	662	280	266
Apples	82·0	660	280	273
Carrots	86·0	527	223	220
Cabbage	88·5	434	184	178

Pancreatine and pancreatine emulsion are sometimes valuable in assisting the digestion, and malt extracts also. All these substances can be looked upon as supplementary to diet, and not as medicines. If cod-liver oil is objectionable, butter should be eaten at every meal, or cream be made use of.

In the event of high temperature supervening, the digestion of meats is greatly interfered with, and the usual diet stands in need of modification, as the nitrogenous matter will, if excessive, embarrass the digestive powers, and prove an encumbrance to the stomach, leading to further complications, which may be avoided by substituting food of a different nature that has not to undergo such a complicated process of absorption and elimination. For this purpose milk and raw eggs are to be chiefly relied on, with beef-tea, soups, jellies, light puddings, toast, biscuits, and farinaceous substances, as arrowroot, or one of the numerous " foods." By this regimen, bearing in mind that the system requires much less nourishment when the body is at rest, an ample dietary is furnished. During bad weather also, when but little exercise can be undertaken, and confinement indoors is called for, diminished diet proves of some service, not only to the comfort of the patient, but to his general condition, and consequently, the local state of disease.

On the cold, dry days, with outdoor exercise, the appetite can be wholly satisfied, with safety and advantage, for it is on these occasions that the "push" is given to nutrition, and any excess in nourishment is more likely to be burnt up or assimilated in the system, to maintain heat, produce force, or counterbalance waste or change.

As the breakfast is not a substantial meal, it will be perceived that the latter part of the day, between noon and 8 p.m., is the period principally occupied by the digestive process, the remaining sixteen hours, therefore, will, in the case of delicate persons, prove to be a great tax on the force and heat-producing powers unless sustained in some way or other. Although it is impossible to lay down rules to apply to every one, a short dietary table for an ordinary case of loss of flesh can be modified to suit the temperament or capacity of any individual, bearing in mind that in many instances "suitable diet" is a matter of experiment, and can be better adopted under medical advice.

Regimen at the Swiss Health Resorts.

7 a.m.—Warm milk, $\frac{1}{2}$ litre; slightly warmed by the vessel being placed in hot water.

8 or 8.30 a.m.—Breakfast : Tea, coffee, or chocolate, bread, butter, honey, porridge. Extras not provided *en pension*— eggs, cold meat, bacon, omelette, &c.

Noon or 1 p.m.—Lunch : Soup, meat, fresh vegetables, sweets, cheese, a glass of red wine, or wine and water ($\frac{1}{2}$ pint).

4 or 4.30 p.m.—Warm milk, $\frac{1}{2}$ pint, with a biscuit, or other light refresment, as tea or chocolate, with bread and butter, or rusks.

6.30 p.m.—Dinner : Soup, fish, or *entrées*, meats, vegetables, sweets, cheese, red wine or wine and water ($\frac{1}{2}$ litre).

Supper not provided *en pension*. Milk, $\frac{1}{2}$ pint, with biscuit, &c., or beef-tea, or some "food" prepared with or without milk.

A glass of milk, with alcohol in some form may also be taken sometimes at 11 a.m., or during the night when medically ordered.

If night-sweats occur, nourishment should be taken at frequent intervals, especially during the night. Stimulants are efficacious at these times—but spirits should always be mixed with milk or egg, or both combined. Their efficiency seems to be increased in this way. Neat spirits as a "*petit verre*" cannot be recommended with much benefit unless there is food in the stomach. Brand's extract of meat and Liebig's, are suitable, also Johnston's fluid beef is an excellent preparation. Veltliner wines drunk with meals will act as a good astringent. A frequent cause of perspiration at night is an excessive quantity of clothes on the bed. The usual eiderdown quilt should not cover the patient if night-sweats are frequent. A flannel night-shirt is of great service on these occasions, and the temperature of the bedroom ought never to rise over fifty-five degrees. Fifty to fifty-five degrees is generally found to be a comfortable temperature in winter.

Exercise, Meals, Sleep, etc.—It is unnecessary to point out the need for individual management in this matter. Much will depend on the state of health and capability for exertion. One rule can, however, be laid down—viz., keep in the open air as much as possible. This will of itself entail a fair amount of movement, but if the state of the lungs does not preclude skating, coasting, tobogganning, or walking ascents—care must be taken in beginning exercise gradually. A state of breathlessness or fatigue must never be permitted, nor must the body be allowed to cool rapidly if perspiring.

Dr. Solly, in writing on Colorado Springs and Manitou, says,—"One of the commonest errors that health seekers make is in matter of exercise. They usually carry it to excess. For instance this is a frequent experience here, for an invalid to call with a note of introduction from his home physician, saying that soon after coming he felt so well he did not think it necessary to present it, but now after some week's residence he finds he is not nearly as well, and came to see if the altitude was not too great for him."

The time available at these altitudes in the depth of winter is somewhat limited for delicate persons, as the sun at this time of the year remains but a short time in the valleys (five or six hours). It is therefore incumbent on those seeking health to make the most of this time, and when in-doors to breathe as much fresh air as can be admitted with comfort. Bedroom windows can be left a little open if the nights are clear, and the heat of the stoves regulated accordingly. In giving an outline of personal hygienic management a description of how the day may be passed will be of some assistance ; but, it is needless to observe that where any pulmonary or other trouble exists, written directions cannot supersede medical supervision.

At 6.30 or 7 a.m. the room should be warmed, and a half-litre of warm milk be brought to the bedside of the patient. After drinking this, an hour's sleep may be had. If a cold bath be prohibited, a rapid sponging of the chest and back, followed by friction, is of great service in keeping the skin in a healthy condition. A fairly vigorous person may have a cold bath, or the chill may be taken off; but precaution must be observed in having the air of the bedroom at this time fresh and warm (not below 55° Fahr.), so that the deep inspirations caused by the shock of cold to the skin shall not take in the used-up bedroom air. A bath or sponging may be tolerated in many cases, provided it takes place *immediately on rising*, while the body is hot. If the skin be allowed to cool down by tardiness in preparing for the bath the water will feel intensely cold. Slight dumb-bell movements may be executed when dressed, throwing the shoulders back and taking deep inspirations. After breakfast the patient should get out for a walk in the sun-light, making a slight ascent if not too short of breath and the state of the lungs permit, returning at 11 a.m. for a glass of warm milk. The afternoons are mostly spent sitting out of doors, skating, coasting, walking, or sleighing, the latter is not to be recommended in very cold weather, without sun. Another half-pint of milk can be drunk in the afternoon, or a little light refreshment such as tea or coffee will do no harm, the patient being generally able to determine if it can be taken with benefit. If he finds it is too much fluid, a less quantity of thin cream might be substituted. Dr. Symes Thompson gives a short piece of advice free from mysticism on the subject of exercise in the Alps (a):—"Those in health need few restraints, but for those with active lung disease sudden exertion on arrival should be discouraged lest it lead to hæmorrhage. If there is active disease or hæmorrhagic tendency or moist sounds in the lung, the patient should sit out in the sun till dry sounds replace the moist ones. He may then walk on the level, or skate, or gently stroll up and down hill, thus causing deep inspirations. Quiet skating can be indulged in by almost all. Tobogganing is more severe, as patients are apt to talk and laugh when walking up the hill. This is very good for the vigorous, as it expands the chest. Lawn tennis is suited only for the strongest, in whom lung disease is quiescent."

The dinner hour varies in different hotels from 5·30 to 7 o'clock; 6·30 p.m. is a very good hour. Food should be taken leisurely, and masticated well. Half an hour's rest before and after meals facilitates digestion. The evenings are spent in various ways. Amateur theatricals are got up at the hotels with concerts, tableaux vivants, balls, &c. There is a good string band at Maloja, and one at the Kulm Hotel, St. Moritz.

The out-door amusements which really enter into a part of the "treatment" at these places are skating, sleighing, tobogganing, and snow-shoeing, besides the ordinary exercise of walking, sitting out in the open air and in shelters, sketching, reading, &c. After breakfast, as soon as the sun appears on the snow a walk is taken, or the skates put

(a) "On the Winter Health Resorts of the Alps." E. Symes Thompson, M.D.

G

on and the ice rink visited. Some sit in shelters or on the
ice watching the skaters, protected from the glaring sun by
smoked-glass spectacles and parasols. Those desirous of
getting as much open air exposure as possible have their
lunch sent to them on the ice or in wooden sheds constructed
for these purposes. After lunch the skates can be changed for
the toboggan, or a sleigh drive taken to some village or place
of interest. A snow-shoeing party may be got up, and a
small excursion made to the inaccessible nooks of the woods
and side valleys, all of which are rendered easy of exploration by the Canadian snow-shoe. Picnics, too, are often
undertaken in this way or by sleigh, combined with a run
down a sloping road on the Canadian sled. The temperature on these occasions may be from 15° to 25° Fahr. ; the
snow and roads are quite dry, and a shawl thrown over the
surface of the crisp dusty snow serves well for reclining or
sitting down.

With respect to the length of time for sleep, the temperature and habits of the individual will have to be considered.
The old dictum of six hours for a man, and seven for a woman
will scarcely commend itself to most people, nor does experience teach us that any definite duration of time for mental
and physical rest can be determined with exactitude. The
intensity of muscular, mental, or nervous exhaustion during
the day will in all cases influence the desire for repose. Cold
also disposes to sleep, as may be witnessed in hybernating
animals. Some human beings also hybernate. It will not be
too much to say that eight or nine hours slumber in winter,
at these cold stations, is near the mark ; remembering that
nothing is to be gained by remaining in bed in a semi-state
of wakefulness after this time is past.

If smoking cannot be entirely given up, the quantity of
tobacco used should be cut down as low as possible. Cigarette
smoking should certainly be discontinued by those whose
lungs are affected, as the habit generally acquired of inhaling

the smoke or passing it through the nasal passages, proves very irritating to the mucous membrane, and is more injurious to the throat and lungs than the same portion of tobacco smoked in a pipe.

CHAPTER VI.

The South West End of the Engadine.

SINCE the development of Davos Platz into an Alpine Winter resort and a rapid growth in size and population, other places have sprung up as having strong claims on our consideration. Wiesen the picturesque, with its ponderous flanks and countless pine trees, is on a fair way to offer health to a limited number who desire quietude and the charms of country life. The milder temperature and extreme calmness give a special character to this resort, which renders it suitable as a change at the end of the winter season for those who have been living at a higher level.

St. Moritz, as before mentioned, attracts many people in winter, and has been well filled for several seasons. Many other small hotels and private apartments have been duly patronised by English people. Dr. Frank Holland has done much to render the place popular.

It is indispensable for the reputation of the high valleys in Switzerland that more stations be developed for the "winter cure," as it is notorious to those who give real attention to the subject that, although these places increase in size, population, and independence, fresh enterprise is needed to keep pace with overcrowding, and to preserve the clean air of the Alpine heights free from contamination. As but few hours can be spent out of the house in winter, it is imperative that hotels, which harbour within their walls a number of delicate individuals in search of health, should be rendered as perfect as possible in a sanitary point of view, Heretofore, the curative properties of these climates have only been developed to a fraction of their full power; and whilst the restorative agent has laid at the closed doors and double windows of over-loaded hotels, patients have been sent home cured in spite of what would seem to be to on-lookers either an under-estimation of the true antidote or the popular disinclination to incur any expense for efficient sanitary arrangements.

A laudable venture was made in the construction on the Maloja plateau of a large Kursaal designed especially for winter residence, fitted with arrangements for the propulsion and extraction of air throughout the building. The Kursaal faces the lake of Sils, between which and the N.E. façade lies a large extent of ground for promenading in summer; and on the S.W. side of the building several acres with ice-rinks, shelters, &c., used mostly in winter, as this is the sunny side. A concert room is semi-detached from the main corridor to guard against any transmission of sound to the interior of the house. This structure is also warmed and ventilated by propulsion and extraction, but in a slightly modified manner to that of the general plan.

An entirely new departure has been taken as respects the drainage of mountain hotels. A plan to check the entry of sewer-gas into the dwelling has been adopted, and the main difficulty in these climates—viz., freezing of the water supply to *cabinets*, precluding the use of syphon traps, has been overcome; each *cabinet, urinoir,* &c., is efficiently trapped, ventilated with the external air, and warmed in winter.

A description of the system of ventilation in use will be given further on.

MALOJA.—The meaning of this word has caused much speculation, and as various constructions have been put on the origin of "mal" and "loggia," the opinions of two Swiss authorities on such subjects will be interesting. Dr. Killias, of Coire, who was kind enough to furnish me with information on the subject, writes as follows :— "The meaning of the word Maloja or Maloggia (locally Malöja or Malöggia; pronounced Malögia) is as difficult of explanation as innumerable other local names. The Rhaetic names are mostly of very old origin, being derived, it is said, from the Etrusk, Keltic, and other antique Rhaetic origins. As regards receiving the apparently Italian or Latin origin, one must be very cautious, much nonsense having been already published thoughtlessly, e.g., the explanation of Celerina as 'celer' oenus. Now there the river Inn does not run any faster than in the neighbourhood. This name has simply been changed in favour of the well-sounding Italian word, the real and true name being Romansch, 'Schlarina,' as found on all old maps, but falsely explained latterly by a Latin etymology. The first mention of the name Maloja I find in a list of references of the Bishopric at Coire (edited between 1290 and 1298, where it is written Malöggia). Campell, in 1572, writes, Malögia, and explains the name after the manner of the scholars of that time, e.g., 'Majores Juliæ,videlicet alpes.' Now the name of Julier has nothing to do with Julius Cæsar, who never came to Rhaetia. Sererhard (1742) speaks of Maloja, and mentions an inn which stood there. No doubt at those times no first-rate lodgings could be had, but in all probability no worse than in the villages round about; therefore, the explanation of 'bad-lodging' is an idle word-play. The root 'mal' has clearly two meanings—I. Latin, in which case it is often found at the end : Via mala, Pass mal, Val mala ; not Malvia, Malpass, Malval. II. There must be quite another word at the bottom, probably Celtic, as there are a great number of names with the 'mal' at the beginning, and where the meaning of 'bad' has no sense at all— Malenco, Maladers, Malans, Mals, Malenz (Vintschgau),

L'HÔTEL-KURSAAL DE LA MALOJA

Malaer; then others not in the Canton of the Grisons, Malbun, to be explained, 'bad good'! Malfrag, Malgera, &c., &c. I cannot say absolutely that the 'mal' is derived from the Celtic 'mael,' a mountain, is right or not, but I decidedly refuse the Latin signification of these cases.

"Some time since a Mr. Pallioppi, of Celerina, compiled a Romansch vocabulary, which ought to have been printed; unfortunately the author died before this could take place, his son, however, occupies himself with the subject, and would be able to give you more exact details."

The Rev. Emil Pallioppi writes:—

"My late father takes the name Maloja from the Celtic, Malögia, Malöggia, Maloja, either taken from the Cymric 'moelang' or 'moelôg,' which means a place rich of hills, as is the case in that part of the Engadine, or synonymous with 'mullach,' top of a mountain or culm; wherefrom also comes the word 'mualach' (môlac) with the signification of road pass. The transitions from u or o to a, and vice versâ, have no difficulty at all, because those three vowels in Celtic are often used one for another (vide Pallioppi). Or the word is composed from 'mala' a hill, mountain, top of a mountain, culm, and 'oiche,' water. In fact is, the small hill or pass on the Lake of Sils, or not far away."

The **MALOJA PLATEAU** (5,941 feet) is situated at the higher extremity of the Upper Engadine, and terminates in the so-called Maloja Pass, where the descent is made into the Bergell Valley. Facing the plateau is the lake of Sils; the largest of the four lakes between Maloja and St. Moritz. In contrast with the brightness of its varying colours, and the clear blue of the sky, are the rugged crags of the higher Alps, clothed below with Juniper, Rhododendrons, Mountain Ash, and Larches. The site is considered by many to be the most romantic and picturesque part of the Engadine.

On the eastern wing of the lake and plateau lie the

Bernina chain : comprising some of the loftiest mountains of
the Grisons and the whole of Switzerland, matted with
glaciers and snow fields to an extent of more than 350 square
miles. The slopes of the Corvatsch, Surlej, and Della
Margna (their peaks elevated to ten and eleven thousand
feet), abut on the lake of Sils, and the eastern side of the
plateau, protecting the lower ground. Opposite and round-
ing towards the north west the rough broadside of Lunghino,
Gravasalvas, and La Grêv are in close proximity. To the
south the Muretto, Dei Rossi, Del Forno, and Salecina, rear
their summits to heights of 10,000 feet, whilst a mile or so
to the south-west is seen the prominent Piz Lizzone and the
serrated crests on the eastern ridge of the Val Bregaglia.

GEOLOGY OF MALOJA. (a) —The immediate neighbourhood of the Kursaal presents some interesting memorials
of glacial and other past times.

To realise the physical facts presented to the eye, it is
necessary to remember the changes which have taken place
in the past period of time that has elapsed since the mountains and valleys were first subjected to the action of ice,
water, and other destructive and formative agents.

The highest authorities agree in the opinion that there
were four glacial periods, interrupted by inter-glacial epochs
of mild and genial conditions. In the Engadine valleys the
mountains show that at the greatest ice-period the glaciers
had a depth of 2,400 feet. At the second ice period the
depth was 1,200 feet, but it is supposed to have lasted longer
than the first. The precipitous character of the lower 300
feet of the mountains in various places, appears to justify
the opinion that, that was the depth of the glaciers during
the third ice period.

Many of the Swiss valleys afford evidence of their

(a) Written for the Author by the late Francis Lloyd, Esq., F.R.G.S.

having been occupied by large and deep lakes. The terraces on the taluses left by the glacier, and on moraine deposits, at the bases of the mountains near Pontresina, Celerina, Samaden, and lower down the valley of the Inn, warrant the supposition that the valleys of the Engadine were, at some far distant time, occupied by one vast lake, 300 feet deep, and that for a long period of time such a lake would have extended from the Maloja to the existing Roseg and Morteratsch glaciers, and probably as far down the Inn valley as Zernetz.

The first glacier which occupied the lake of Sils had its origin between Campfèr and Silvaplana, their being every appearance of Piz Albana and Piz Surlej having been united at some period, and there being unquestionable evidence of great glaciers having flowed, from about that point in opposite directions. That flowing southwards, which for distinction, I will call the Sils glacier, was fed by the mountains right and left and by tributary glaciers from Val Julier, Val Fex, and Val Fedoz. The two latter join the Sils glacier near the Maloja on the south-west side of the valley. In consequence the lower part of the mountains on the west side are much shattered and very precipitous, whilst those on the south-east side are in the main rounded.

It seems highly probable that in the remote past Piz Lunghino and Piz Margna, the two last mountains, right and left at the Maloja were united by a mountain chain. Of this there are now only comparatively small remains, the highest path being only 300 feet above the lake. This chain is broken through in four places. The first cleft is at the side of Piz Lunghino and is very narrow. The second cleft is on the west side of the mass of rock on which the Château (Belvedere) stands while on the east side of the same there is a much wider and deeper cleft (Val d'Enfer) through which the later and smaller glacier probably flowed. It is a remarkable feature that there is a deep transverse cleft in the rock-

mass on which the Château stands. There is a similar but
smaller transverse cleft through the first rock-mass. The
sides of both show much ice wear and seem to justify the
opinion that they were cut by the ice which flowed from the
basin between Piz Materdell and Piz Lunghino. The fourth
rock mass extends to near the Kulm and between it and the
base of Salecina there is a wide cleft, or rather depression, in
the deepest part of which flows the Ordlegna river from the
Muretto valley.

The road to the Kulm (at telegraph post No. 219) is cut
through a moraine which curves round on the right to the
top of the fourth rock-mass. On the left it also curves
behind, but at some distance from the Kulm hotel. Thus far
it appears to be strictly moraine, but further on as far as the
rocky knob which commands a view of the bridge over the
Ordlegna there is a considerable deposit of large granite
boulders distinguished by large quartz crystals lying on the
rock. The rocky knob commands a good view of the whole
of the much-curved moraine, which it may be noticed is
backed by a rocky bridge on which are many similar boulders.
Within the curve may be seen another smaller but well-
formed moraine on which stands four châlets and there is
within the last a very small moraine-like ridge, grass-
grown. The largest of these moraines is cut through at one
part and the other two appear to have been acted on by
water. Close to the smallest there is a low-lying wet
pasture.

Around the Kursaal Hotel there are several small hills,
grass-grown, of the height of 40 feet. Below Gresta at the
S.W. corner of the lake there is a small hill composed
entirely of remarkably fine sand, and between that and the
Hotel there is a mass of glacier-worn rock—a good specimen
of furrowing and striation.

I proceed now to attempt an explanation of the striking
features I have pointed out. The Forno glacier had its

source in an enormous ice-field at a great elevation. At the mouth of its valley it met the Sils glacier. Side by side they flowed to the Val Bregaglia. No mortal eye saw, nor can imagination picture, an ice-river nearly three miles broad and 2,400 feet deep, falling as a cascade into the valley which is now more than 1,000 feet below. As it ploughed its channel between the mountains we must not expect to find any of its moraine in the Sils or Forno valleys. We see only the shattered mountains and the furrowed, striated, and smoothed surfaces, as evidence of its power.

The glacier of the second ice-period although only half the thickness, must have continued the wear and tear. Assuming that in the third ice-period the glaciers were 300 feet thick they would have carried before them most of the moraines left by the glacier which preceded them.

The perpendicular rock which supports the south angle of the Château shows one half of an ancient *moulin* about 3 feet in diameter.

The moraines described as lying near the Kulm probably belong to the third ice-period, and were the lateral moraines of the Forno glacier which must have met the Sils glacier here, and was therefore forced to bend round to reach the Val Bregaglia. The granite boulders are of a kind that is not found in the Sils Valley, the mountains bounding the latter being mostly of talcose or micaceous slate. The granite boulders abound in the Val Muretto and also encroach on the Sils Valley proper.

On the supposition that the Sils and other Engadine lakes were at some period of a depth of 300 feet more than now, the waters must have been dammed up, probably by moraine of the second and third ice-periods, deposited on the rock-chain above the Val Bregaglia. It has already been stated that the part next to Lunghino is 300 feet high. If we suppose moraine to have been left at the same height on the other parts right across the valley, a lake would have

had the depth stated. As time elapsed the upper part of
the barrier would have been sapped, and the lake thereby
reduced in depth. This process would have been repeated
until the depth was lowered to that of the col above the present level of the lake of Sils. I have stated that the well-
defined moraine near the Kulm is in one part severed, and
that close to the Kulm Hotel there is a low-lying wet
pasture. Both are probably the result of a rush of water
from the lake to the Val Bregaglia. There is scarcely
room for doubt that the hills around the Kursaal are
composed of drift gravel, and are really lake sediment.
They consist of rounded pebbles, small stones and sand :
there being in places seams of sand only. This material was no doubt brought down by the rivers from the
tributary valleys, it having all the appearance of river gravel,
and not the character of moraine matter, in which large
rounded or worn boulders are rarely if ever absent. These
hills were evidently at one time a continuous bed. They are
40 feet high measured from the ground at the base of the
Kursaal. The channels dividing the hills coincide with
existing water courses from the surrounding mountains. The
largest of these channels starts from the base of the Château
hill and runs directly towards the site of the Hotel. The
depth of the deposit implies a deep lake existing a long
period of time. The small bed of remarkably fine sand I
attribute to an eddy at that part of the lake which admitted
of the fine sand being deposited from the water in which it
was suspended.

———————

The thalwind or valley-wind, which blows in every Swiss
valley, although stronger than in the Davos district, seldom
exceeds a force of one degree Beaufort scale, is by no means
insupportable, and dies away in force and frequence when
snow covers the adjacent regions. At this season, and when

the lake of Sils freezes, greater calm prevails, and the locality frequently partakes of that Alpine stillness and sunshine which allows the most delicate individuals to be exposed to a low temperature without feeling the sensation of cold. Preceding and during a fall of snow in the mountains of Switzerland, storms are not uncommon, alternating with lulls and blasts of wind. Commonly these proceed from a southerly or south-westerly direction, and bring with them the peculiar changes in the atmosphere, which have given rise to a special name for southerly currents of air, viz., föhn wind. This föhn, which is considered the pernicious wind for invalids is rarely perceived at the S.W. end of the Engadine, its peculiar characteristics being tempered by the high altitude, and by passage over a portion of the Bernina snow fields and glaciers. In this way the temperature of a southerly current is greatly reduced, and the commencement of the föhn shorn of its oppression and sultriness. Rapid rises in temperature at the onset of a southerly wind during winter are never so pronounced at this altitude as at 5,000 feet.

One of the most important features of Alpine climate— and one which unquestionably acts as an adjuvant in the chemical process of the natural and healthy change taking place in the blood discs—is the amount and duration of sunlight. Were it not for the transparent air and powerful sun the high Alps would be quite unsuited to delicate people. By reason of the direct chemical action of light, and the increased time afforded for out-door amusement and exercise which a longer day admits of, an additional hour's sunlight during the short days of December and January is an inconceivable advantage, both for the mind as well as the body. The inequality, in this respect, of Davos, Wiesen, and St. Moritz, with the Maloja, is due, as may be supposed, to the configuration of the surrounding mountains. If a slice could be taken off the Jacobshorn at Davos, and the Stulsergrat, at

Wiesen, the sun on the shortest day would flood those districts long before half-past ten, as is the case now.

During the months of December and January the Maloja has more than one hour's sunshine longer than Davos, Wiesen, or St. Moritz. In respect of duration of sunlight, Davos Dörfli and Pontresina still exceed the Maloja, but do not enjoy the *late sunset of the latter.* At page 125 the reader will find the sun-hours given for the various health-resorts during the winter months. Sunrise and sunset take place on the first day of January as follows :—

			Sunrise.	Sunset.
Maloja	...	...	9.35	3.45 P.M.
Wiesen	...	...	10.35	3.45 ,,
(a) Pontresina	....	...	8.30	3.10 ,,
(a) St. Moritz	...	...	10·0	3.5 ,,
Davos Dörfli	...	...	8.35	3.17 ,,
Davos Platz	....	...	10.3	3.0 ,,
(b) Andermatt	...	....	11.45	3.15 ,,

The necessity for exposure to Alpine solar rays in the treatment of various complaints, is well illustrated by cases of anæmia among some of the domestics employed in hotels at these levels. Attracted by the rumours of cure in anæmia they seek employment at some health station, but being principally confined to kitchens and shady places their anæmia rests with but little improvement. The case with nurse-girls is different as they very soon gain a healthy tint, being constantly in the open air.

Another advantageous feature of this part of the Engadine, is the vast extent of level ground which admits of exercise being taken without fatigue. After the breathing becomes freer and there is a tendency for the chest to expand,

(a) Taken from the houses on the main road of the village (Dr. Ludwig's "Ober-Engadin").

(b) Reported by Dr. Schmid.

tobogganing and gentle ascents can be undertaken. For the more vigorous a descent into the Bergell Valley on a sled is one of the finest runs in Switzerland. Skating of course, can be indulged in by those who are equal to the pastime ; a good rink has been constructed, whilst the lake of Sils is but 300 yards from the Hotel.

The climate can be described in the general terms of all Swiss high level stations, viz., a stimulating, bracing, tonic, and cold climate ; increasing the power of the heart and other muscles, expanding the chest and exciting the nervous and glandular systems into healthy activity. The Maloja shows a more equable temperature than Davos, and more wind ; but the winter season lasts a little longer, which is a great advantage to those who make a change to lower levels, as the weather is generally more settled the further spring is advanced. The rapid rises in temperature caused by the föhn in mid-winter are by no means so frequent or so high as in the narrower situations. It is not pretended that any part of the Upper Engadine is, on the whole, as calm as the Davos valley ; the latter is so by reason of its being narrower (limiting the sunshine, and favouring stagnation of air) but what one health resort loses on one hand it gains on the other.

Omitting the occasional storms that sweep over the whole of the Alps, the gentle and bracing currents of air which move everwhere in the Upper Engadine, although not suited to serious forms of illness, are pleasant by reason of their dryness, and prove of the greatest benefit to those who have sufficient resisting power. In all the narrow valleys it must not be forgotten that the dead-calm days and burning sun are usually precursors of the föhn wind, which has peculiarities of so much importance to delicate persons.

During mid-winter a dip of over 3,200 feet can be made in an hour and a-half to Promontogno, and Chiavenna (1,090 feet) can be reached in less than three hours, if the upper level appears to disagree. This facility in being able to quit the highlands without a prolonged, and in cases of illness, a dangerous journey from any other Alpine station, cannot be over-rated. There are instances in which any climate, either cold or temperate, does not always effect the salubrious change anticipated; it then becomes a question of the utmost importance to an invalid, how to escape without incurring risk. In the present instance the dangers of moving, even in the depth of winter, are at a minimum. A drive to the lake of Como requires less time than it takes to reach Davos from Landquart.

The soil and vegetation of the Maloja are of the usual character found at these altitudes, a thin, rich, dark incrustation of the former, rapidly absorbing moisture. The bed of the plateau (formed by an ancient recession of the lake of Sils) consists of peaty mould, resting on a layer of gravel. Up to a height of 600 feet above the level ground the slopes on the eastern side are covered with Larches and Mountain Ash. The western side is rugged, large boulders and crags being interspersed with patches of nutritious short grass. On the plateau itself, and the little hills adjoining, are dwarf firs (*Pinus montana*), Heath, Juniper, and Cranberry bushes.

The coldest part of the Engadine is near Bevers; the temperature in winter becomes a little higher as we approach the Maloja, owing to the fact that its south-west end terminates in an abrupt fall of 1,200 feet (*a*) into the extensive valley of Bergell, which after passing Promontogno terminates near Chiavenna in Italy, 18½ miles distant; the Italian frontier being 12½ miles from the Maloja. To the proximity of a channel at this end for milder air to mount and replace the denser and colder strata, the deviation in temperature may be attributed.

(*a*) The precipitous descent is cut by a road with sixteen zig-zags.

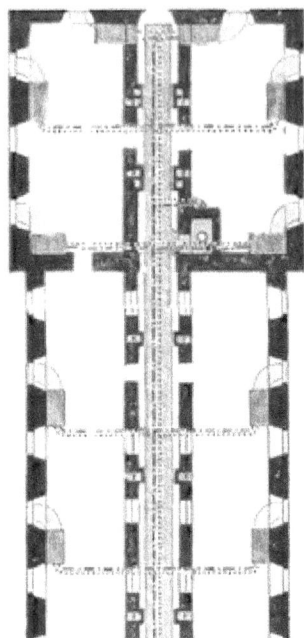

Batteries de Chauffe.

Entrée de l'air extérieur.

Sortie de l'air vicié.

À BRUXELLES.

Conduit de fumée.

Conduit de vapeur.

Conduit d'eau de condensation.

air of a bedroom once every two or three hours. In the concert, smoking, and billiard rooms, and public salons, a necessary number of conduits are provided for renewing the atmosphere at least once every two hours.

Each delivery tube is fitted with a sliding valve which can be adjusted at pleasure according to a degree of heat desirable for comfort.

To allow of the escape of used-up air, two tubes of exit are placed in a chamber, one under the bed for the winter ventilation, another near the ceiling, used in summer to exhaust from the warmer stratum of air. These exit tubes are in connection with a casing surrounding the main flue of the *chaudières*, that acts as an extraction shaft for the foul air of the central portion of the building. The kitchen chimney acts in this manner for another part of the house and together with the flue of the confectionery and that of the bakehouse, placed in the basement, aspirate the air of the rooms into double chimneys. As the upward motion in these extraction shafts would scarcely be forcible and rapid enough during summer time to effectually withdraw the air from so many chambers, and as perchance some air might find its way back again to the upper stories, a contrivance with steam tubes is fixed in the space beneath the roof to again heat and thereby promote the ascent of foul air and so accelerate its escape externally.

Every corridor, salon, bed-room, and *cabinet* in the house is ventilated in this fashion, each chamber having its own separate ventilation by three channels—one inlet and two outlets (one for the escape of warm impure air, the other for the cooler air near the floor). By careful management and adjustment of the valves, the rooms can be regulated to a wholesome temperature. Even in summer, if the *batteries* are not heated, extraction will go on as usual ; and the windows may be opened without interfering in any way with the general ventilation of the house.

☐ Entrée de l'air exterieur.

☐ Air chauffé.

☐ Air vicié.

▬▬ Conduit de vapeur.

.... Eau Condencée.

H   Ventilation d'hiver.

Ǝ   Ventilation d'été.

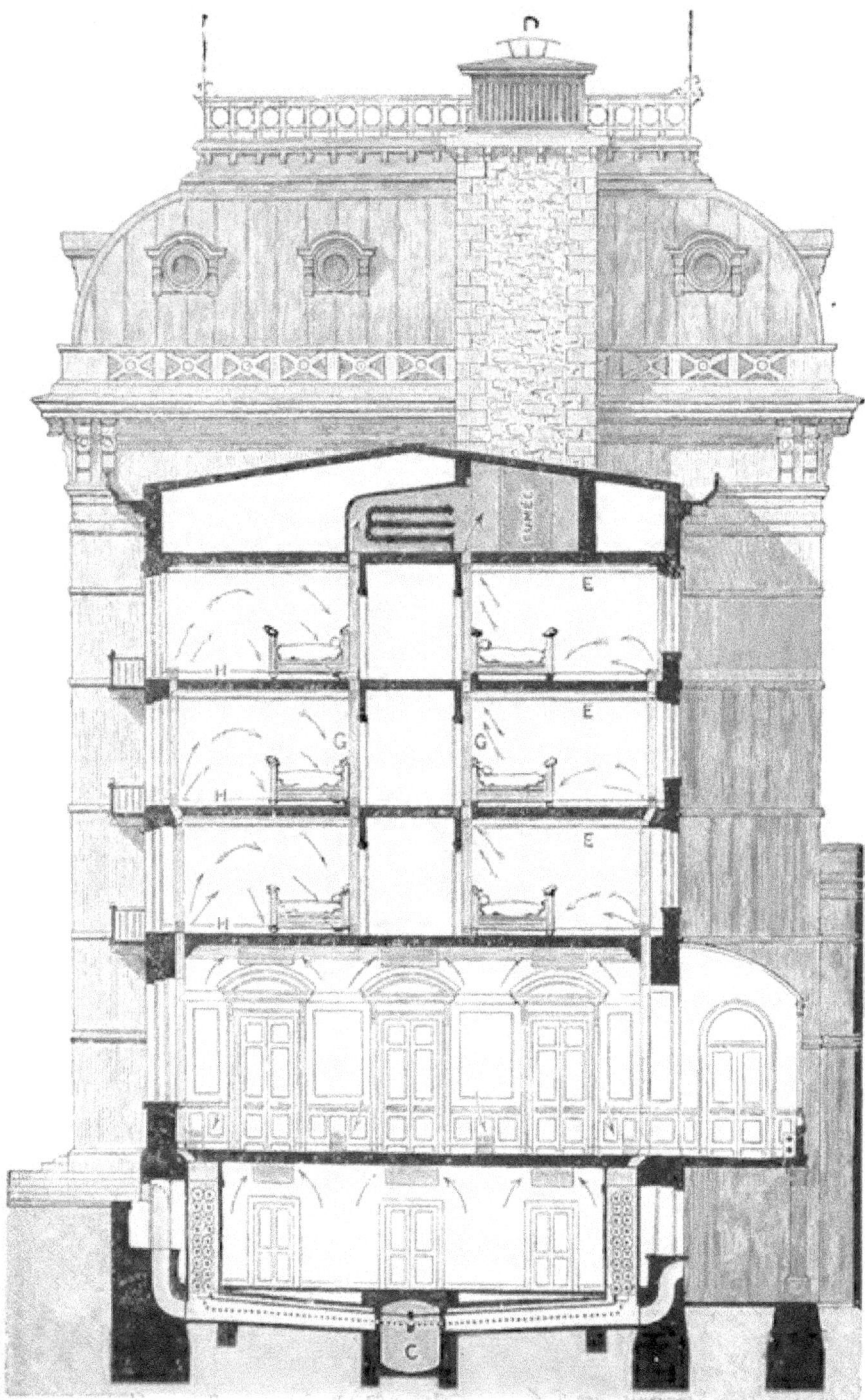

Maçonnerie.

Enveloppes des batteries.

Appareils des chauffage.

Attaches et supports.

Vase de saturation.

Prise d'air.

Conduit de vapeur.

Purgeur.

Robinet.

Tuyau d'ecoulement de l'eau.

Trappe montée à charnières.

Registres pour modérer.

Valve de melange.

Conduits d'air.

Idem aux étages.

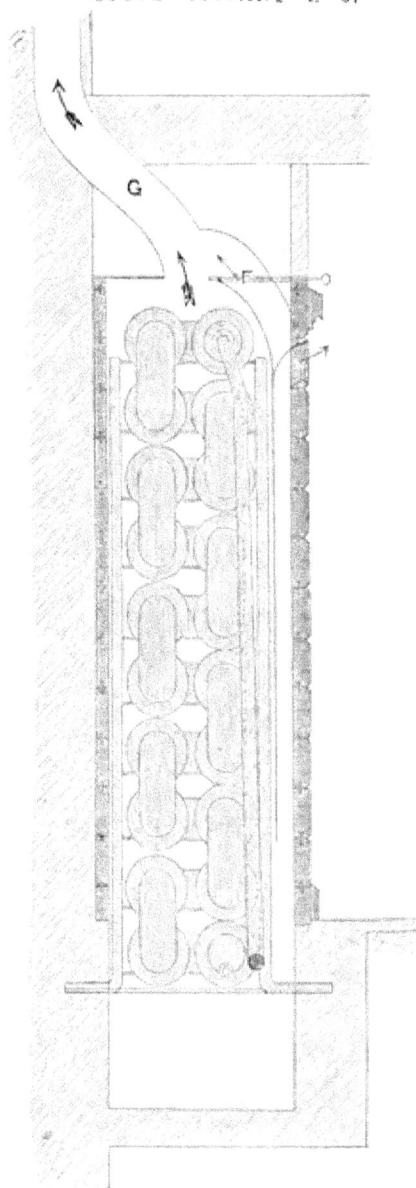

G

COUPE SUIVANT K. L.

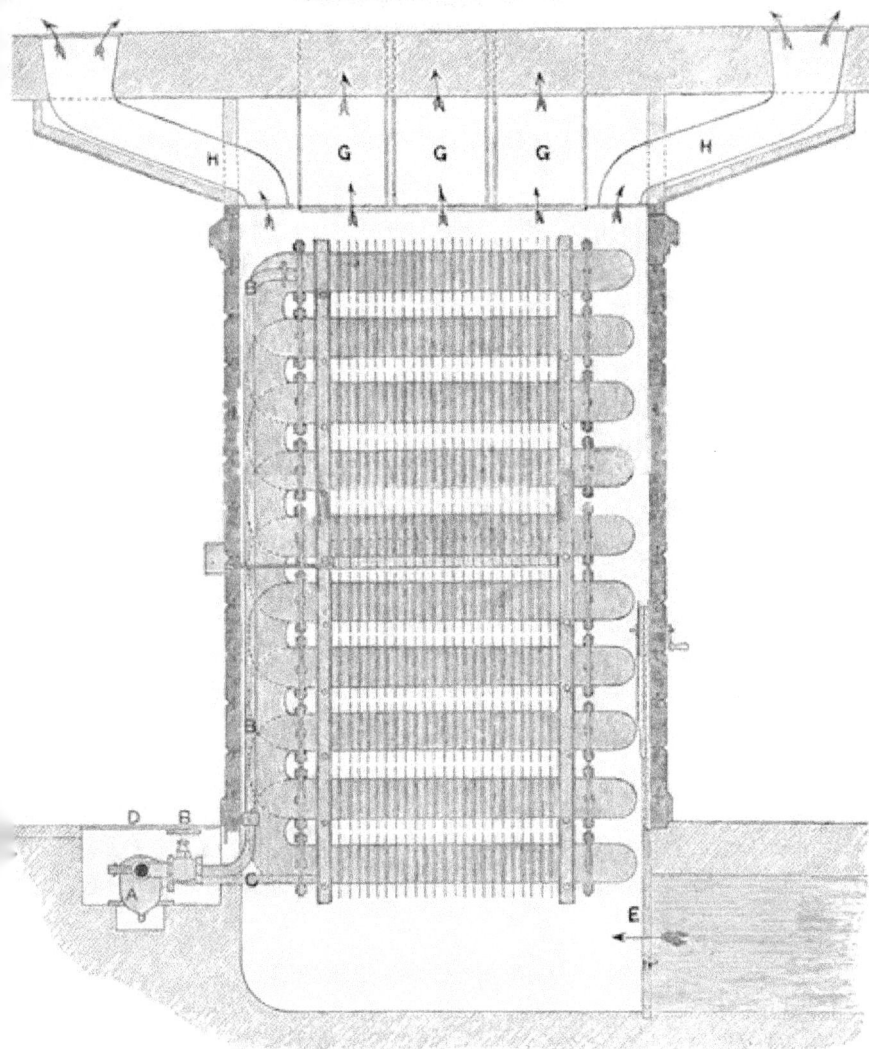

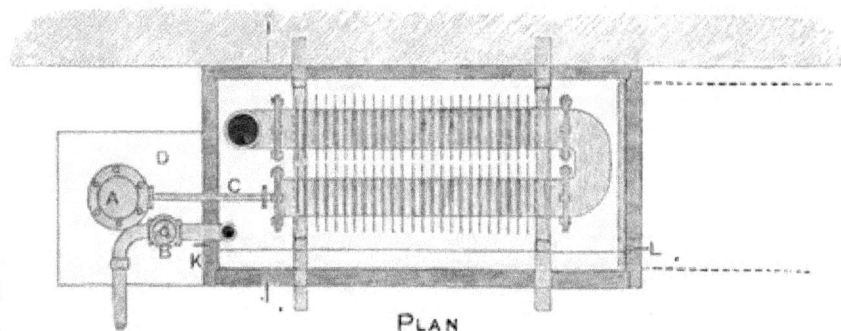

PLAN

Such an elaborate system, if well-attended to and carried
out under intelligent supervision, admits the mountain air
within the house in as pure a condition as is practicable,
moistened to compensate for the raised temperature.

In no sense does there appear to be a contra-indication of
the summer Alpine climate for persons with chest complaints,
that is with some of the early phases of phthisis, not those
with hopeless advanced symptoms, and who are in a feeble
irritable condition, with high fever, and incapacity for gentle
exercises. The determination of suitable cases for the high
altitudes, both in summer and winter can be roughly gauged
by the amount of physical power conserved, taken in conjunction with the duration and apparent rate with which
consolidation or softening is advancing. The slow scrofulous
forms and fibroid conditions receive undoubted benefit, but
these must not be complicated with secondary lesions.

The length of time needed for a permanent recovery of
lost strength or health, has ever been considered too short to
carry with it a firm re-establishment of exhausted energy. A
summer month spent in the Engadine—with assiduous
regard to diet, exercise, sleep, and medical treatment—should,
in almost every instance, be prolonged to six weeks *at least ;*
longer if time permitted. The lengthened stay would in the
long run, be an economy of time as far as health was concerned, and prove a greater satisfaction to doctors and patients
alike. Force of a more lasting and durable nature would be
accumulated, the system settle down in a more staple condition, than if the treatment be suddenly relinquished to
return to old occupations and former habits.

It is true that physiological changes take place with great
rapidity and completeness. A good indication of such being
the case is the demand made by nature for more fuel, shown
by the increased appetite : by the skin being braced up with
or without the stimulus of baths : the kidneys and the whole
glandular system being awakened into activity : blood

changes quickened and amplified : and above all, the percep-
tion that the capacity for mental and physical exertion is
extended ; but if a departure is made immediately improvement of the health is attained and strength begins to be felt,
some disappointment may eventually be experienced, and
the climate be discredited by the transient effects of too short
a stay to allow the frame to be seasoned in the new conditions, or undergo the peculiar alteration known as acclimatisation. This is especially noticeable in anæmia and
chlorosis, where the blood changes take place rather rapidly,
and a hue of health is acquired in a few weeks leading to an
erroneous conclusion that a permanent cure is effected
whereas a further time is needed for its completeness. In
chronic cases, either of catarrhs (pulmonary, gastric, or
uterine), phthisis, anæmia, some abdominal affections, and
even dyspepsia, a medical mind is aware of the value of
prudent management after the disappearance of symptoms.
A variety of cases have improved in a most remarkable
manner. It is also undoubtedly true that in these high
altitudes some rapidly advancing symptoms have been
apparently arrested when scarcely any hope of a return to
health could be held out by physicians.

Patients with the following diseases should, however,
refrain from seeking health at high elevations :—

1. Diseases of the brain, heart, or large vessels.
2. Tendency to articular rheumatism.
3. Kidney diseases (during winter).
4. Acute inflammations of throat or larynx.
5. Some diseases of bladder or prostate.
6. Nervous or vascular excitement.
7. Pronounced hysteria.
8. Persons somewhat advanced in years should not visit
the mountains, unless the circulatory system is sound; nor
those suffering from extreme emphysema of the lungs.

It can be quite understood that no definite rules can be

laid down in a book, as to the exact state of health which would be benefited or aggravated by dwelling at high levels, one can only point out the class of cases for which mountain climate is suitable or contra-indicated. An absolute decision in individual instances is only to be arrived at correctly by those with clinical experience of the effects of these potent climates on various bodily conditions.

It has been supposed for a long time, that nearly all forms of valvular disease of the heart contra-indicate a resort to high levels, either for health or pleasure. Some correspondence took place in the *Medical Press and Circular* on this point.

I have met with two or three cases of mitral insufficiency, which required treatment and then went on as usual. One case, also, of an ancient rupture of a valve (caused by violent running), is extremely interesting. On the occasion of each visit made to Maloja (6,000 feet) breathlessness and palpitation demand immediate treatment which is continued off and on for the first month, when everything goes on well again. In spite of singular irregularity of the pulse, plenty of exercise is taken, skating, walking and tobogganing, &c., the Maloja pass being descended (800 feet) and remounted.

I can neither verify nor refute Dr. Oertel's theory that true affections of the heart can be remedied by graduated mountain climbing ; but my experience has been that the general health and strength of a few of these sort of cases have improved greatly : at the same time it is right for me to say that I have not directed special observation to this subject in the way that Dr. Oertel has ; and secondly, that I have always found graduated exercise in the mountains not only salutary for my cases of pulmonary troubles, but absolutely indispensable, for nearly every condition of ill-health which comes under my notice.

The varieties of pulmonary troubles for which high altitude stations are most beneficial are the following :—

1. Cases of predisposition to lung troubles, either from frequent attacks of catarrh or bronchitis, or from hereditary taint. (a)

2. Imperfect development of the thorax, with shortness of breath on exertion.

3. Hæmorrhage from the lungs with or without physical signs of disease.

4. Incomplete recovery from pneumonia and from pleurisy, where the lung does not expand after removal or absorption of any effused fluid.

5. Chronic tubercular phthisis with moderate pyrexia, but with a good area of sound and serviceable lung tissue in reserve.

6. Spasmodic asthma without cardiac complications or emphysema.

7. Primary cavities in the apex, where the disease is circumscribed and inactive.

The following pulmonary conditions are unsuitable for residence in high climates: Acute phthisis, and laryngeal phthisis; catarrhal phthisis (epithelial pneumonia of Sir Andrew Clark); suspicion of pulmonary aneurisms; chronic bronchitis or bronchiectasis; marked emphysema of lungs; phthisis with basic or secondary cavities (primary cavities in the axilla have the tendency to heal up and contract, but, according to Dr. Ewart, secondary cavities in dorso-axillary regions never heal, and the sternal region and base of the lungs are unfavourable positions for the healing of primary cavities. These observations were made by Dr. Ewart at the Brompton Hospital, on a large number of lungs, and are now

(a) There are many persons who cannot go through an English winter without "catching cold" on the slightest exposure, change of weather, sitting in a draught, or getting wet. These "colds" end in cough, and are among the chief causes in establishing an insiduous consolidation or catarrh of a lung.

so important regarding the prognosis in phthisis at high altitudes, as to justify repetition).

All cases of phthisis complicated with diarrhœa, pyrexia, albuminuria, profuse hamoptysis, or anæmia where the hæmoglobin is down to 50 per cent., or the white corpuscles of the blood number 2 per square, by the hæmacytometer of Hayem.

Primary cavities in the apices of the lungs, where there is an insufficient residue of healthy lung substance left for breathing with comfort.

Excitable natures, prone to neuralgia, pyrexia, restlessness, variable pulse, generally over 120 beats, and irregular sleeping powers, termed by some writers the erethric constitution, should seek health elsewhere than at high level stations.

The foregoing definitions of the states of disease which do well in high lands, and the reverse, pretty clearly represent the class of cases suitable for treatment.

It is true that some foreign writers have attempted to elaborate on these conditions, and have endeavoured to lay down more minute details and rules for guidance; they have, however, generally succeeded in depicting a state of health which alone should preclude the advisability of the patient leaving home for any climate whatever.

High altitude treatment has long quitted the arena of theory; what is wanted now is practical familiarity with the forms of disease which have been seen to rally towards cure from winter residence. It is, however, to be clearly understood that there must be some vestiges remaining, of resistance, stamina, recuperative energy, vigour, or whatever one chooses to term this undefinable entity, with all invalids (whatever the physical signs of disease may be) who are to derive benefit from the regions of snow and ice. Once this point is passed the invalid's existence is by no means to be envied, it is fortunate if he can get away to softer climes.

Europe, moreover, does not offer the supreme advantage of high altitude with a moderate temperature, especially in a class of cases, where weakness and depression preclude a certain amount of gentle exercise, and where the heart is weak and irritable or the circulation so languid as to give rise to chilliness and cold livid extremities. As far as our knowledge goes at present such climates are to be found only across the Atlantic Ocean.

In the following cases treatment has been omitted, but it may be mentioned that the practice carried out varied somewhat with each patient, and consisted of graduated exercise, with suitable clothing and diet, plenty of fresh air, dry cupping, medical rubbing, and counter irritation. Milk, cod-liver oil, and medicines when circumstances demanded it ; drugs were, however, avoided as much as possible.

The cases are taken as they occurred in my note book of pulmonary complaints of the past winter ; there are no omissions of unsatisfactory results.

No 1.—Æt. 26. Congestion of right apex, dulness, feeble breath sounds, vocal resonance increased, and slight recurring crepitation. Pulse, 64 ; temperature, 97·2. Weight, 8 stone 7 lbs. Slight cough ; no expectoration.

Result—Increased expansion of chest, right side 1 centimetre, left side ½ centimetre ; pulse, 60 ; temperature, 97 to 98·4. No abnormal physical signs. Marked gain in strength, appetite good. Weight, 9 stone 6 lbs.

No. 2.—Æt. 29. Congestion of right apex, slight dulness, with harsh expiratory murmur, hæmoptysis on three or four occasions during the previous twelvemonths, no cough nor expectoration. Weight, 9 stone 3 lbs. Pulse, 108.

Result.—No hæmoptysis during the winter, great gain in strength and capacity for exercise—skating, walking. Voice power increased ; is now able to sing, although previous attempts during

(a) Extract from a paper on the Climate of the Swiss Alps, by the Author. Read at the International Medical Congress, held at Washington, U.S.A., September, 1887.

H a

the last three or four years have always resulted in failure and hoarseness. Weight, 9 stone 5¾ lbs. Pulse, 100.

No. 3.—Æt. 24. Cough and hæmoptysis two years ago when in the United States. The appetite failed at that time, and there was great loss of flesh with night perspirations. After spending three months in Europe he has undergone much improvement. On examination moist râles were heard over the apex of the left lung with pleuritic crackling and slight dulness. On the right side jerky breathing was heard underneath the clavicle. Yellow expectoration with morning cough. Temperature, 98 to 100. Weight, 9 stone 13 lbs. Bacilli in sputum. Pulse, 100 to 116.

Result.—After summer residence dry sounds replaced moist sounds. The expectoration was about the same. Weight, 10 stone. After winter residence the weight was 9 stone 13 lbs., it had been up to 10 stone 5 lbs., but was reduced by indiscreet exercise, as all abnormal physical signs and general health had improved. Pulse, 84 to 96 ; temperature, 98·4. Bacilli in sputum undiminished in number.

No. 4.—Æt. 49. Consolidation of the superior lobe of the left lung. Commenced with slight hæmoptysis a few months ago. There was no loss of flesh nor night perspirations. Cough frequent, with expectoration of yellow heavy mucus, occasionally streaked with blood. Pulse, 88 to 104 ; temperature, 98·4 to 99. Weight, 9 stone 10 lbs. Bacilli in sputum.

Result.—Strength greatly increased after the winter's residence, and expectoration lessened. Weight, 9 stone 9 lbs. Pulse, 90 to 120. This patient unfortunately died in Italy during the subsequent spring from an attack of pericarditis.

No. 5.—Æt. 28. Left front of chest immobile, portions of three ribs have been excised for empyema. Respiratory murmur inaudible over the left lung, where there was a cavity in the apex communicating with pleura. Harsh breathing in right apex, extending to second rib, otherwise the right lung was normal. Pulse, 96 ; temperature, 98·4. Morning cough, and occasionally violent fits of coughing after meals. Breathlessness on exertion ; appetite good ; sleeps well. Weight, 10 stone 8 lbs.

Result.—General improvement. Recommended a sea voyage

and then to return to Maloja for a second winter. Weight, 10 stone 12 lbs. Pulse 92.

No. 6.—Æt. 15. Illness commenced in the spring of 1882 with slight cough, and hæmorrhage on two occasions, malaise and anæmia. The winter of 1882-3 was spent at Bournemouth, 1883-4 Algiers, 1884-5 Tunbridge Wells and the Riviera. There was always a little cough and yellow expectoration. The sputum was tinged with blood at Alassio in April, 1885, and the temperature at that time ranged from 99·4 to 101. Came to the Engadine in July, 1885, when the following notes were made :—Mucous membrane of mouth and fauces anæmic. Dulness over the whole of left lung, with numerous dry and moist râles. Rhonchi heard over the lower lobes of right lung. There was retraction of the cardiac lobe, and displacement of the heart, with flattening beneath the clavicle extending to fourth rib. Fatigued easily, and breathless on slight exertion. Pulse, 116 ; temperature, 98·8 to 99. Bacilli in sputum.

Result.—There was gradual improvement during the summer of 1885. On the 19th November, 1885, she weighed 7 stone 5 lbs. ; at the end of March, 1886, 7 stone 10 lbs. ; March, 1887, 8 stone 13 lbs. The aspect of this patient was greatly changed for the better ; she was able to skate, toboggan, and mount 500 feet of Maloja Pass without fatigue. The left side of the chest was considerably contracted, but gave slight signs of filling out again. A moderate sized cavity was dry and contracting. There was still a little cough with nummular expectoration in the mornings. Bacilli were found in large numbers. Pulse 80 to 100. Temperature normal. After a third winter the pulse had fallen to 72 and 92 ; a slight cooing rhonchus was still heard over the left lung, the cavity of which had cicatrised and closed to small proportions ; ordinary exercise could be taken without fatigue, and the patient felt quite well. Weight, 9 stone 11 lbs. ; gained 34 lbs in 30 months.

No. 7.—Æt. 26. Contracted pneumonia in 1877, 1883, and 1885. Absorption of inflammatory products was incomplete in right apex. Expiratory murmur was harsh, and prolonged over the whole of the right lung. Weight, 11 stone 2½ lbs.

Result.—Respiration normal. The right infra-scapular region still remained dull on percussion, but the fringes of the lungs in front gave signs of compensatory emphysema. Weight, 11 stone 7 lbs.

No. 8.—Æt. 20. Insiduous consolidation of right apex. Dulness over right subclavicular region with augmented vocal vibrations, rough and prolonged vesicular murmur, bronchophony and morning cough with muco-purulent expectoration, loss of appetite and strength, with night perspirations. Temperature 98·4. Bacilli in sputum.

Result.—The ten months—June to March—were passed between Davos and Maloja, at the end of this time there was a disappearance of all serious signs. Slight flattening could be discerned beneath the right clavicles, where the dulness and excessive vocal vibration were almost imperceptible ; bronchophony insignificant ; no cough nor expectoration. Weight, 10 stone 10 lbs.

No. 9.—Æt. 35. Loss of weight generally for eighteen months with dyspepsia. Loss of voice after speaking for 10 or 15 minutes. Perspires a little at night. There was harsh vesicular murmur over the right apex and feeble breathing in the right base, but no râles or other adventitious sound. Pulse 84. Weight, 11 stone 10 lbs.

Result.—A slight harshness in expiration alone perceptible over the right lung. Voice stronger and appetite good. There was no dyspepsia, and the strength was greatly increased. Weight, 12 stone 2½ lbs. Pulse 80.

No. 10.—Æt. 27. Subject always to winter colds, which end in cough. Dulness over the apex of right lung, with bronchophony and harsh expiration. No râles or crepitations. Becomes breathless and fatigued after exercise. Pulse 76 (sitting), 100 (standing). Weight, 11 stone 5½ lbs.

Result.—He could mount a 1,000 feet without breathlessness or fatigue. All abnormal signs had disappeared. Pulse 76 (sitting), 92 (standing). Weight, 11 stone 6 lbs.

No. 11.—Æt. 32. Hæmoptysis for four or five years, loss of flesh, and bad family history. Never much expectoration nor high temperature. There were pains complained of over the apex of the right lung, where there was high-pitched jerky respiration, but no râles. Liver enlarged. Pulse 80. Weight, 8 stone 8¼ lbs.

Result.—After 23 days residence in the month of February the body weight increased to 8 stone 12 lbs., but there was no change in the physical signs. Pulse 72. A little hæmoptysis occurred on one occasion attributable solely to indiscretion while mounting the pass.

No. 12.—Æ. 45. An Anglo-Indian, on two years' sick leave. There was dulness under left clavicle, with a small area of increased vocal resonance. No râles of any kind. Pulse 80. Loss of muscular power.

Result.—This gentleman diminished in weight whilst in Europe from 13 stone 4 lb., to 12 stone 4½ lb. with great advantage. He could take much exercise, descending the Maloja Pass (800 feet) on a toboggan, and mounting again on foot, sometimes twice daily. No abnormal signs remained. Pulse 72.

No. 13.—Æt. 20. Convalescent from typhoid fever and pneumonia. Pulse 100 (sitting), 112 (standing). Weight, 11 stone 3¼ lbs.

Result.—After one month's residence, in February and part of March, there was considerable gain in strength and respiratory power. Pulse 92 (sitting), 100 (standing). Weight, 11 stone 8 lbs.

14.—Æt. 40. Was at Davos about six years ago with catarrh of right apex, which was completely cured by three or four months' residence. There was now shortness of breath and dulness over both apices, with harsh expiratory sound and dry rhonchus. Very slight increase of vocal resonance anywhere, but limited expansion of upper part of chest. Pulse 80 (sitting), 85 (standing). Bacilli in sputum. Temperature 98·4.

Result.—Beyond an increase in the capacity for exercise, and a diminution of dry rhonchi, there was nothing to note.

No. 15.—Æt. 19. Slight cough for four years. No expectoration nor loss of flesh. There were imperfect respiratory sounds in both apices of the lung, especially marked on the left side with deficient entry of air over the whole of left dorsal region, which was somewhat dull on percussion. No râles. Pule 92 ; temperature 99. Weight, 8 stone 6 lbs.

Result.—Normal entry of air was manifest in both apices. The dulness over the left lung was greatly lessened. Temperature 98·4, pulse 78. Weight, 8 stone 4 lbs.

No 16.—Æt. 40. Obtained great benefit from wintering at Davos six years ago. During the last twelve months constant colds have ended in a cough with yellow expectoration, sometimes streaked with blood. There was bronchial breathing and dulness over the apex of the left lung with marked vocal resonance and bronchophony back and front. The signs indicated congestion around, and in the neighbourhood of a cavity, which had cicatrised and contracted six years ago. An occasional dry rhonchus was heard in the left base. Pulse 100 (sitting), 120 (standing), temperature, 97 ; respiration 24.

Result.—The time spent in Maloja (one month at the end of winter) was insufficient to produce any very marked improvement in the physical signs. The cough, however, was much diminished and appetite improved. Pule 84 (sitting), (100 standing) ; temperature 97·8 ; respiration 20.

No. 17.—Æt. 23. Prolonged, expiratory murmurs in both apices, especially perceptible on the right side with increased vocal resonance and fremitus. There was slight cough with muco-purulent expectoration containing bacilli. Pulse 100 (sitting), 124 (standing), intermits on deep inspiration ; temperature 98·2. Weight, 11 stone 3 lbs.

Result.—Entire disappearance of all abnormal signs in the chest. He was able to take ordinary exercise without any fatigue. Pulse 92 (sitting), 100 (standing), no intermissions on deep inspiration. There still remained a slight expectoration, in which bacilli could be distinguished. Weight, 11 stone 13 lbs.

No. 18.—Æt. 37. Consolidation of the whole of left lung, dry rhonchus, whispering sounds and crackling over the upper lobe, with shrinking and uncovering of the heart, a troublesome cough, dyspnœa and loss of appetite. Respiration 26 ; the temperature varied from 97 to 101 ; pulse 116 to 125. Bacilli. Weight, 8 stone 3 lbs.

Result.—This case was confined to bed a good deal during the whole winter, on account of dyspnœa. Apparent improvement was, however, noticeable by a slight expansion of the left chest and marked expansion of the right, taken by cyrtometric tracings. Pulse 92 to 100 ; temperature 97 to 99·4. Weight, 7 stone 12 lbs. Bacilli undiminished.

No. 19.—Æt. 23. Illness commenced three years ago with a cough and loss of flesh. There was dulness and imperfect entry of air in both apices, more marked on the left side, where crackling was heard. The mucous membrane of palate and fauces was pale and anæmic. Cough, and yellow expectoration containing bacilli. Pulse 116 (sitting), 124 (standing) ; 140 (after examination of the chest). Weight, 8 stone 11 lbs.

Result.—Cough almost gone and expectoration much lessened, air entered more freely in both apices and left base, but the dulness had not diminished. Pulse 88 (sitting), 100 (standing) ; temperature 97 to 99. Weight, 9 stone 11¼ lbs.

No. 20.—Æt. 43. Was cured fifteen years ago of a "breaking down of lung tissue" in left apex by one winter's residence at Davos and several years afterwards spent in South Africa. Recently, after a bad attack of bronchitis in Germany, a return of old symptoms appeared. There was considerable flattening of the right subclavian region, with impaired resonance, moist rhonchi were heard over the whole of the lung, especially clear in the interscapular space. Vocal fremitus was not increased, nor was absolute dulness perceived anywhere. In both apices there was a harsh expiratory murmur on deep inspiration. Pulse 120 (sitting) ; temperature varied little from the normal. Bacilli in sputum. Occasional cough with muco-purulent expectoration. Weight, 10 stone.

Result.—During the portion of the winter (3 months) spent at Maloja he had two or three attacks of feverishness in the afternoons, the temperature rising to 101. At the end of three months dry rhonchus was heard in the right lung, and increased resonance in both lungs could be made out. There was no gain in strength and expiratory power. Pulse 112 (sitting), 120 (standing) ; temperature 100 (taken at 5 p.m.). Weight, 10 stone ½ lb. This patient had to leave on business before completing a whole winter's residence.

No. 21.—Æt. 23. Enjoyed good health up to four or five months ago, when hæmoptysis occurred. The chest presented signs of congestion of left apex : dulness, increased vocal fremitus, slight bronchophony, and harsh expiratory murmur. Pulse 76 (sitting), 100 (standing), intermits on deep inspiration ; temperature 98. Weight, 8 stone 12¾ lbs.

Result.—All dulness of chest disappeared, the expiratory murmur was still a little harsh on both sides of the chest in the subclavicular regions, otherwise there were no abnormal signs. Pulse 68 (sitting), 92 (standing) ; temperature 98. Weight, 9 stone 7 lbs.

No. 22.—Æt. 22. Extremely bad family history (mother and three sisters died of phthisis). There was imperfect expansion of the chest, with cooing rhonchus in left apex, extending over the cardiac lobe, and heard plainly between the scapulæ. Dry click in right apex, slight expectoration, coloured once or twice with blood. Pulse 88 (sitting), 104 (standing) ; température 99. Weight, 9 stone 1½ lbs.

Result.—With the exception of a slight increase of vocal resonance over the right apex, there was no dulness nor abnormal sounds anywhere. Pulse 76 (sitting), 80 (standing) ; temperature ranges from 96·8 to 98·1. Weight, 10 stone 1½ lbs. Was able to take plenty of exercise, skating, snow-shoeing, tobogganing, &c. After a third winter's residence, spent more as a precautionary measure than a necessity, the pulse was 68 and 72, and the body temperatures ranged a little higher, but still kept below normal, viz.—97·8 to 98·1. Weight, 9 stone 13 lbs. No bacilli had ever been discovered in the expectoration, which, however, was insignificant in quantity.

No. 23.—Æt. 19. Was in an anæmic and debilitated condition from over study. There were no signs of any damage to lungs, but the expansion of the upper portions of the chest was insufficient, and the family history bad. Pulse 108 (sitting), 120 (standing) ; temperature 98·4; Weight, 9 stone 8 lbs.

Result.—Complete disappearance of anæmia, and much gain in strength, takes a fair amount of exercise. Pulse 92 to 100. Weight, 9 stone 6½ lbs. Expansion of chest increased one centimètre.

No. 24.—Æt. 42. Cough commenced six years ago; first hæmorrhage occurred in May, 1887; second hæmorrhage in August, 1887, about one pint; was very weak after this, and cough increased, with thick, yellow expectoration. No appetite; great weakness and dyspnœa, but no night sweats. Extensive consolidation of right apex, with softening, consolidation of left apex. Pulse 92 to 114. Weight, 8 stone 13 lbs.

Result.—Gained strength gradually during the winter, cough and expectoration much lessened. Weight, 9 stone 1½ lbs. Expansion of chest 1 centimètre each side. Pulse 72 to 96. Bacilli in sputum apparently undiminished in number.

No. 25.—Æt. 26. Winter cough for four years. First hæmorrhage twelve months ago; hoarseness and cough, with muco-purulent expectoration, containing bacilli; no fever nor perspirations. Consolidation and catarrh of the upper third of right lung, dulness extending to the third intercostal space. Pulse 76 to 96. Weight, 8 stone 10¾ lbs.

Result.—Dulness over the upper part of right lung, and mucous râles much lessened; gain in muscular, respiratory, and cardiac power. Pulse 68 to 84. Weight, 9 stone 3½ lbs. Bacilli diminished in number.

No. 26.—Æt. 45. Hæmoptysis in 1886 and 1887; family history good, winter cough and expectoration for two years. Dulness over both apices, extending to third intercostal space on the right side, with numerous fine crepitations; marked dyspnœa and immobility of the chest. Pulse 92 and 108. Weight, 12 stone 2 lbs.

Result.—Dyspnœa lessened, and capability for exercise increased; cough and expectoration but slightly diminished; more resonance of chest, with expansion of four centimètres on the left side and three centimètres on the right. Weight, 12 stone 9½ lbs.; bacilli undiminished; pulse 92 to 104.

No. 27.—Æt. 17. Frequent attacks of bronchial catarrh during winter, necessitating urgent treatment with inhalations, liniments, medicines, and careful nursing. Chest rather narrow and pigeon-shaped, emphysematous respiration, but no consolidation

nor râles ; pale complexion.   Pulse 92 to 84, no dyspnœa.
Weight, 8 stone 8¼ lbs.

Result.—No catarrh during the winter ; expansion of chest,
one centimètre on the right side and 1 centimètre on the left.
Pulse 72 to 84 ; healthy appearance.   Weight, 8 stone 13 lbs.

No. 28.—Æt. 25.   Cough, loss of flesh, perspirations, and
hæmoptysis.   Consolidation of right apex, extending to third intercostal space.   Pulse 100 to 116 ; temperature 98·4   Weight,
8 stone 1 lb.   Anæmic.

Result.—Dulness over right apex almost disappeared, respiration more extended and stronger, muscular power increased, and
absence of anæmia.   Pulse 80 and 88.   Weight, 8 stone 4 lbs.

No. 29.—Æt. 40.   Pleuritic effusion eight years ago ; cough
ending in hæmoptysis twelve months ago.   Dulness over right
apex, extending to third space in front and to fifth, dorsal vertebra
behind.   Pulse 104 to 116.   Temperature 99·2.   Dyspeptic symptoms ; muco-purulent thick, and profuse expectoration, containing
bacilli.   Weight, 7 stone 11 lbs.

Result.—There remained slightly higher pitched percussion
sound beneath right clavicle, but no absolute dulness nor râles ;
interscapular spaces normal ; expectoration diminished, still containing bacilli ;   great gain in respiratory and muscular power.
Pulse 84 to 96 ; temperature 98·6.   Weight 8 stone.

No. 30.—Æt. 17.   Always delicate ; eight brothers and sisters
have died of various complaints, principally lung diseases ; consolidation of the upper half of left lung after pleurisy, no great shortness of breath, anæmia and debility ; cough troublesome, with
viscid muco-purulent expectoration containing bacilli, night perspirations with occasional fever ; no entry of air in axillary region
on the left side, on a level with the 8th rib.   A few moist râles
heard in the lower part of left lung (which is immobile), and over
the third intercostal space on right side.   Pulse 92 in bed, 96, 108,
120, and 128 at various times, difference in sitting and standing
position generally 8 beats.   Weight, 8 stone 3¾ lbs.

Result.—Was able to take a fair amount of out-door exercise,
skating and tobogganing, cough about the same ; expectoration
thicker and less in quantity, bacilli remain ; air can be perceived to

enter over the whole of the left lung, but in a feeble and imperfect manner; the left side of the' chest now responds to inspiration. Pulse 104 and 108; gain in expansion of chest 2 centimètres on the left side, and one on the right. Weight, 8 stone 10½ lbs.

No. 31.—Æt. 25. Hæmoptysis about two months ago, without any marked physical signs of disease. Slight dulness and dry click under clavicle : cooing râles in supra-spinous space on the right side. The chest measures 1 centimètre less on this side. Bacilli in sputum. Pulse 116 to 114. Temperature 100·2, night perspirations and loss of flesh with slight cough. Takes a great deal of exercise. Weight, 7 stone 1¼ lbs.

Result.—Contracted a lobular pneumonia in mid-winter through indiscreet exertion and exposure; made a good recovery, and regained former condition in six weeks. Physical signs remain about the same, but no fever nor night sweats.

No. 32.—Æt. 37. Pleurisy and slight hæmoptysis 12 months ago. Takes cold easily and sleeps badly. Sore throat and catarrh : depressed in strength and spirits. Imperfect entry of air in the upper portions of both lungs. The right side measures 42 centimètres and the left 38. Complains of pain below left clavicle. Impaired resonance on both sides of the chest but no dulness. Weight, 9 stone 5¾ lbs. Pulse 88.

Result.—Resonance over the whole of the chest, and freer entry of air. Expansion 1 centimètre on the right side and 2 centimètres on the left. No depression of spirits. Takes plenty of exercise. Pulse 72. Weight, 9 stone 7 lb.

No. 33.—Æt. 35. Hæmoptysis, hoarseness, and sore throat, with pallor, loss of flesh, and dyspepsia. Dulness below right clavicle, with imperfect entry of air and dry click, no râles anywhere. Imperfect entry of air in upper portion of right lung. The right side measures 44 centimètres, the left 45. Bacilli in sputum. Weight, 10 stone 12¾ lb. Pulse 84 and 88.

Result.—Great gain in flesh and strength. Dulness almost disappeared with normal entry of air over the whole of lung. Increase in expansion of chest, the right side measures 46 centimètres, the left 47. Bacilli still remain in sputum. Weight, 11 stone 6 lbs.

ABSTRACT OF PHTHISICAL CASES.

No.	Age.	Duration of residence.	Weight gained	Weight lost.	Diminution in pulse rate.	Remarks.
1	26	4½ months	13 lbs.	—	4	Remarkable improvement.
2	29	4½ ,,	2¾ lbs.	—	8	Great improvement.
3	24	7½ ,,	—	—	16	Slight improvement.
4	49	7½ ,,	—	1 lb.	none	,, ,,
5	28	4½ ,,	4 lbs.	—	4	General improvement.
6	15	30 ,,	34 lbs.	—	34	Remarkable improvement
7	26	4½ ,,	4½ lbs.	—	4	Great improvement.
8	20	5 ,,	not noted	—	not noted	,, ,,
9	35	4½ ,,	4½ lbs.	—	4	,, ,,
10	27	4½ ,,	½ lb.	—	0	All signs of disease disappeared.
11	32	23 days	4¾ lbs.	—	8	Marked improvement.
12	45	4½ months	—	15¼ lbs.	8	All signs of disease gone.
13	20	1 ,,	4¾ lbs.	—	8	Marked improvement.
14	40	3 ,,	not noted	—	0	General improvement.
15	19	4 ,,	—	2 lbs.	14	Slightly better.
16	40	1½ ,,	not noted	—	16	Marked improvement.
17	23	4½ ,,	10 lbs.	—	8	All signs of disease gone, with the exception of bacilli in sputum.
18	37	4 ,,	—	5 lbs.	16	No improvement.
19	23	4½ ,,	14¼ lbs.	—	28	Remarkable improvement.
20	43	3 ,,	¾ lbs.	—	8	Slight improvement.
21	23	5¼ ,,	8¼ lbs.	—	8	Great improvement.
22	22	4½ ,,	14 lbs.	—	12	Remarkable improvement.
23	19	2 ,,	—	1½ lbs.	16	Improvement.
24	42	4½ ,,	2½ lbs.	—	20	Marked improvement.
25	26	4½ ,,	5¼ lbs.	—	8	,, ,,
26	45	4½ ,,	7½ lbs.	—	—	,, ,,
27	17	4 ,,	4¾ lbs.	—	—	,, ,,
28	25	3½ ,,	3 lbs.	—	20	,, ,,
29	40	3 ,,	3 lbs.	—	20	,, ,,
30	17	2½ ,,	6¾ lbs.	—	4	,, ,,
31	25	2½ ,,	—	—	8	,, ,,
32	37	2½ ,,	1¼ lbs.	—	18	,, ,,
33	35	1½ ,,	7¼ lbs.	—	—	Remarkable improvement.

NUMBER OF HOURS OF POSSIBLE SUNSHINE DURING WINTER.

	Maloja.†	Wiesen.	Davos Platz.	Davos Dorfli.	St. Moritz.*
1 Nov....	$7\frac{1}{6}$ hours	... $7\frac{1}{3}$ hours	... $7\frac{1}{3}$ hours	... $7\frac{3}{4}$ hours...	— hours
15 ,, ...	$6\frac{3}{4}$,,	... — ,,	... — ,,	... $7\frac{1}{4}$,,	... 6 ,,
1 Dec. ...	$6\frac{1}{2}$,,	... $5\frac{1}{4}$,,	.. $5\frac{1}{4}$,,	... $6\frac{1}{12}$,,	... $5\frac{1}{4}$,,
15 ,, ...	$6\frac{1}{4}$,,	... $5\frac{1}{10}$,,	... $5\frac{1}{12}$,,	... $6\frac{3}{4}$,,	... $5\frac{1}{12}$,,
1 Jan....	$6\frac{1}{10}$,,	... $5\frac{1}{6}$,,	... 5 ,,	.. $6\frac{2}{3}$,,	... $5\frac{1}{12}$,,
15 ,, ...	$6\frac{1}{3}$,,	... $5\frac{5}{6}$,,	... $5\frac{1}{2}$,,	... 7 ,,	... $5\frac{1}{3}$,,
1 Feb....	$6\frac{5}{6}$,,	... $7\frac{1}{4}$,,	... $6\frac{1}{3}$,,	... $7\frac{1}{3}$,,	... $7\frac{1}{12}$,,
15 ,, ...	$7\frac{1}{8}$,,	... $7\frac{2}{3}$,,	... $7\frac{3}{4}$,,	... $7\frac{3}{4}$,,	... $8\frac{1}{12}$,,

* Das Oberengadin (Dr. J. M. Ludwig).

† These observations were taken from one spot on the Maloja (The Kursaal); if the châlets or grounds of the hotel had been included, a very much larger duration of sunshine would be recorded.

MEAN DAILY TEMPERATURE AT 7 A.M., GIVEN IN DEGREES CENTIGRADE.

		Nov.	Dec.	Jan.	Feb.	Mean for the whole winter '83-84.
*Maloja	(6,000 ft.)	—3·2	—6·3	—6·5	—7·9	—5·9
†Wiesen	(4,771 ft.)	—1·5	—5·3	—3·3	—3·1	—3·3
†Davos	(5,105 ft.)	—3·7	—7·1	—5·9	—5·5	—5·5
†Andermatt	(4,738 ft.)	—3·4	—8·1	—5·5	—5·0	—5·5

* Calculated from the observations of M. Kuoni.

† Furnished to the author by Professor Billwiller, Director of the Meteorological Stations.

Davos.

Climate. — Stimulating, bracing, and exciting, but in a less degree than the Engadine.

Position.—Situated on rising ground in the valley itself. The mountain screen ranges from 3,000 feet to 5,000 feet. The mist which sometimes covers the bed of the valley in the early morning, is soon dissipated by the sun, but the usual haze and cloud of smoke seen over villages and small towns is visible unless moved by wind.

Proximity to Glaciers.—Scaletta, 8 miles to S.E. Silvretta, 12 miles N.E. by E.

Strongest Winds.—S. and S.W. (föhn) depressing to invalids.

Snow Winds.—South and south-west winds, as at Wiesen.

Thalwind (valley wind).—Thalwind is sometimes felt between 2 and 3 p.m. from the N.E.
On the whole the Davos valley is extremely calm, but suffers from the föhn wind.

Wiesen.

Less exciting, and, although very dry and restorative, more sedative than any of the higher stations.

Wiesen is located on the hillside, the Landwasser being about 1,000 feet below the houses. The mountain screen ranges from 3,000 to 5,000 feet above. There is in fine weather no morning nor evening mist; a constant but imperceptible current of cold air travels down the declivities. Notwithstanding that this motion of the atmosphere is unfelt, it is sufficient to obviate any tendency to stagnation.

Scaletta, 13 miles to the E. (Small glacier.) Silvretta glacier, 20 miles to N.E.

S. and S.W. (föhn), depressing to invalids.

South and south-west currents, know locally by the name of "föhn."

This wind, which blows in every Swiss valley, is uncertain, as the village is situated far above the bed of the gorge, and consequently out of the zone of commotion of air, caused by the descending cold currents which converge and flow down the gorges and ravines.

ST. MORITZ.

Stimulating, bracing, and exciting, with strong electric conditions of the air and excess of ozone.

The Kulm Hotel is about 300 feet above the lake. A thin mist occasionally hangs over the lake in the morning, but below the level of the dwellings. The aspect is more open than Davos, there is a little more wind, but not sufficient to become disadvantageous in winter.

Extensive Bernina glaciers to S. and S.W. Piz Bernina is 9 miles distant.

S. and S.W. Generally not of a low temperature, but sometimes chilling to delicate persons.

South and south-west wind, much less of the föhn character than at Davos and Wiesen. It is only during the snow-melting that this S.W. wind in the Upper Engadine resembles the true föhn.

No regular valley wind, blowing from a definite direction, independently of the upper current. Nevertheless, the wind rises regularly in the course of the day, and is unpleasant from either the north or south (Mr. Waters' observations).

MALOJA.

Highly stimulating, bracing, and exciting, with strong electric conditions of the atmosphere, and excess of ozone.

Situated at the S.W. end of the Upper Engadine. Long sun-duration in midwinter.

No morning nor evening mist. The Kursaal is placed on the western side of the plateau, and in this situation escapes most of the valley draught (thalwind). Extremely picturesque scenery at this part of the Engadine. Mountain screen, 3,000 to 6,000 feet.

Partially surrounded by glaciers. Bernina glaciers, W.; Murtel, 6 miles; Fedoz, Forno, and Albigna, 4 to 6 miles distant, on the S.W. and S.

N. and N.W., dry, cold, and pungent, but not dangerous nor productive of the nervous phenomena associated with the föhn.

S. and S.W. wind. The sultry and depressing characters of the föhn are extremely rare, probably from the influence of the surrounding glaciers, altitude, and spread of the valley.

Blows from the N.E. principally on the eastern side of the plateau, in a line with the Bergell valley, and is mostly noticed on the Kulm, about one mile distant from the hotel. Although this wind is much stronger than in the Davos valley, it is endurable even in the depth of winter, and has a healthy tonic effect on suitable cases.

THE JOURNEY FROM ENGLAND.

THE quickest route from London to the Engadine during winter is by Dover, Calais, Boulogne, Amiens, Ternier, Laon, Rheims, Belfort, Bâle, and Coire.

The South-Eastern Railway Company's steamers from Folkstone to Boulogne are comfortable boats, and the luggage by this route is taken greater care of, and not so roughly handled as between Dover and Calais, and there is a certainty of always having a large steamer to cross in. Arrangements are now made that a train leaves Charing Cross or Cannon Street at 9.40 a.m. Lunch can be taken quietly at Boulogne, avoiding the ordinary rush for this meal at Calais.

Viâ Dover and Calais, a start is made 11 a.m. from London. Lunch can be obtained at Calais at 3.5 p.m., dinner at Tergnier about 7 p.m., and Bâle reached at 6.15 a.m., where breakfast is taken at the railway station. From Bâle the journey is continued, after breakfast, either to Como by the St. Gothard line, and from thence to Colico by steamer, then by train to Chiavenna ; or leaving Bâle at 7.20 a.m. Coire is reached at 1.10 p.m. (A stay of a few hours can be made at Zürich, if desired, and Coire reached later.)

From Coire the diligence leaves in the early morning for Maloja, viâ the Julier Pass and Silvaplana.

From Chiavenna the diligence leaves at 8.10 a.m., arriving at Maloja 2.45 p.m.

It is advisable for delicate persons to break the journey either at Bâle or Coire, and those who are unable to obtain any sleep in a train would do well to alight at Bâle and go to bed ; otherwise a telegram might be sent to Coire, requesting a room to be warmed, as the air will feel chilly in autumn after sitting still in the train. Should the weather be fine, a few days' stay can be made with benefit previous to mounting. A private conveyance should then be taken for a short journey to Tiefenkasten or Mühlen, spending the night at one of these places, and going on by diligence the next day, when Maloja is reached at 6.30 p.m.

If the route by the St. Gothard railway be chosen (*viâ* Como, Colico, Chiavenna, Promontogno) the train leaves Bâle at 7.15 a.m. for Como, where the night is spent, after which the boat is taken to Colico, and train to Chiavenna, where another rest is made ; or an agreeable stay can be made at Promontogno, situated in a picturesque part of Val Bregaglia, and only four hours from Maloja (ascent).

METEOROLOGICAL OBSERVATIONS

FROM 1883 TO 1888.

METEOROLOGICAL NOTES.

THE meteorology of these high climates is now pretty well understood by those who give attention to the subject, but as the following observations have been made by me personally, I append the summaries for the five years at Maloja.

Although these climates are known to be very dry I would observe that hygrometric observations, although very important in the study of climate, cannot be taken with precision at the lower temperatures without the aid of the chemical hygrometer; it is, therefore, doubtful if the calculations of "weight of moisture," and "drying power of air," are absolutely exact, but they approach as near correctness as possible with the means at our present command. I am in accord with my friend, Mr. A. W. Waters, who has had long experience of Alpine climate, that there is a point where the wet and dry bulbs fail to indicate accurately the humidity of the air. Saussure's hair hygrometer is the most accurate for ordinary observation; it was used by me for the winters 1886-7 and 1887-8.

It may be noticed that the "minimum" temperatures which as a rule occur about 7 or 8 a.m.) are often considerably below the "Dew Point" of 9 a.m. on the same day and

3 p.m. of the day previous, without any fall of moisture
occurring. As a partial explanation for this it may be
inferred that in clear weather the upper strata of the atmosphere contain but an infinitesimal amount of watery vapour,
the air having passed over extensive areas of land, and its
moisture having been condensed by the frozen peaks which
impede the course of aërial currents towards this part of
Switzerland, in which case evaporation from the snow-
covering of the country slowly supplies the lower stratum of
air with moisture, evaporation from the snow diminishing
with the fall in temperature, and although the "Dew Point"
throughout the day and in the evening is generally much
higher than the "minimum" of the night no deposit of
moisture takes place when the temperature falls below the
"Dew Point" as the invisible vapour is quickly dissipated
into space, diffusion ensuing speedily with the lessened
barometric pressure at these heights. That these climates
contain an extremely small quantity of watery vapour in the
air is a well-established fact, but the precise measurement
presents a little difficulty. To illustrate the excessive dryness
in the depth of winter the mean weight of moisture at noon
in 10 cubic feet of air was but 9·1 grains for January; even
this is probably too high a calculation, as the result was
arrived at by Apjohn's formula, which is not quite suitable
for low temperatures. At Kew during the same month
the mean, as given to me by G. M. Whipple, Esq., Superin-
tendent of the Observatory, was 25 grains.

To the absence of moisture suspended in the air may be
ascribed the capability of wild animals to support the cold in
these high regions, and human beings the changes in temperature. It is by no means rare for individuals to quit a room
and proceed out of doors with a difference of 50°, and
occasionally 60° Fahr., without feeling any pressing need of
gloves or extra clothing. On the bright days of mid-winter,
surrounded by snow and ice, the new visitor marvels that the

low temperature is unperceived, and that the sun's rays carry with them the heat of summer. Combined with dryness the cold restrains the life and development of micro-organisms in the external air; ordinary catarrh is almost unknown, except . in badly ventilated houses, where a " cold in the head," or a " sore throat " seems liable to be communicated to others. It can be demonstrated that a drier layer of air surrounds animals and is inhaled by them in these climates, than in Egypt (which has generally been considered the driest air for consumptive patients). By respiration a rise in the temperature of the air takes place, and whatever the temperature inhaled may be, the exhaled air approaches that of the blood. On the other hand the atmosphere in immediate contact with the skin and between garments becomes warmed, consequently what would be expressed by a " relative " humidity of 90 per cent. at 26° Fahr. (the mean temperature for the winters 1884-5) becomes 8 per cent. near the temperature of the lungs, and 13 per cent. in contact with the body, if we calculate the air underneath the clothing at 80° Fahr.

In Egypt the " relative " humidity for the month of June, 1884, was 73, February 70, this has a drying power at the temperature of the body (say 98·6 Fahr.) of 160 and 156 grs. per 10 cubic feet, which is seen to be actually a less dry air when taken into the lungs than that of an Alpine height. Although I have mentioned the "*relative*" humidity for contrast, this observation in the comparison of cold climates with other places, requires a further calculation to afford a definite idea of the hygrometric state of the atmosphere, for with variation in the temperature the standard varies, and what may be a damp climate at 90 per cent. in one locality, is a dry climate in another with regard to animals, for at Cairo 90 per cent. relative humidity represents in January 1884, 37 grains of vapour in 10 cubic feet, whereas at Maloja for the same month but 14 grains are suspended in

the air at 90 per cent.   Therefore, in a medical aspect there
can be no hesitation in saying that the *"absolute"* humidity
should on no account be neglected, the drying power of the
air being given in grains of vapour, then some estimate of
evaporation from the lungs can be made.

SUMMARY OF METEOROLOGICAL OBSERVATIONS TAKEN AT MALOJA DURING WINTER.—By A. TUCKER WISE, M.D.

Date.	Temperature at 9 a.m.	Minimum temperature.	Cloud, 0 to 10.	Wind force.	Solar thermometer. (Fahr.)	Maximum temperature.	Snow. (Centimetres.)	Weight of Moisture in 10 cubic feet of air.	Drying power of the air.	Temperature at noon.
November, 1883	26·4	−11	4	0·7	130	44	56	14	6	30·5
December, „	19·3	−18	4·8	1	124	45	30	11	5	25
January, 1884	19·3	−20	2·8	1·3	129	41	26	11	5	25·7
February, „	17·3	−22	4·3	0·7	143	38	34	12	4	26·7
	20·6	−22	4	0·9	143	45	146	12	5	26·9

MONTH.	9 A.M.		NOON.				3 P.M.				Force of Wind.	Solar Radiation.	Snow.
	Temperature. (Fahr.)	Cloud.	Temperature. (Fahr.)	Cloud.	Weight of Moisture in 10 cubic feet of air.	Drying power of air in grains.	Temperature. (Fahr.)	Maximum. (Fahr.)	Minimum. (Fahr.)	Cloud.		(Fahr.)	(Ctm.)
November, 1884	21·9	2·2	32·3	2·2	16·2	9·8	31·4	53·8	8	4·3	0·9	115	6·5
December, „	22·3	5·6	27	4	12·3	5	27	42	−0·5	5	0·9	100	45
January, 1885	10·9	3·6	19·8	3·3	9·1	2·6	21·7	36·5	−6·5	3·6	0·6	91	59
February, „	20·8	4·2	31·6	4·1	14·1	6·6	32	47·5	−10	4·3	0·7	140	114
March, „	24·7	4·3	32·8	3·8	14·4	7·5	33·4	48	4	5·2	1·5	143	45
	29·7	4	28·7	3·5	13·2	6·3	29·1	53·8	−10	4·5	0·9	143	269·5

MONTH.	AT 9 A.M.			AT NOON.			AT 3 P.M.							
	Dry Bulb.	Wet Bulb.	Cloud. Amount 1 to 10.	Dry Bulb.	Wet Bulb.	Cloud. Amount 1 to 10.	Dry Bulb.	Wet Bulb.	Cloud. Amount 1 to 10.	Force of Wind	Solar Radiation.	Minimum.	Rain Gauge.	Snow.
	Fahr.	Fahr.		Fahr.	Fahr.		Fahr.	Fahr.			Centgr.	Fahr.	m.m.	ctm.
November, 1885	29·8	28·5	7	33·5	31·2	6	32·9	30·9	5	0·5	43·5	11	13·5	12
December, ,,	21·4	20·3	3	27	23·7	3	26·8	24·1	3	1·5	41	0·5	8·7	9
January, 1886	16·2	15·4	5·5	22·4	20·5	5	23	21·2	4·7	1	37·5	−6·5		162·5
February, ,,	10·5	9·8	3	23·2	20·8	3	24·2	22	4	0·5	46	−12		29·5
March, ,,	15·7	14·3	3	26·4	22·9	3	26·5	23·6	4	1	51	−14		32
	18·7	17·6	4·3	26·5	23·8	4	26·6	24·5	4·1	0·9	51	−14	22·2	245

Date.	Temperature at 9 a.m.	Minimum temperature.	Cloud, 0 to 10.	Wind force.	Highest reading of Solar thermometer.	Maximum temperature.	Snow.	Percentage of relative humidity.	Weight of moisture in 10 cubic feet of air.	Drying power of the air.	Mean temperature at noon.	Barometric Range.
1886.					centigrade.		centimetres.		grns.	millimetre		
November	23	5	3	1	35	34	16·5	62	12	7	30	600 mm. to 620 mm.
December	18	-5	3	1	39	37	291	80	12	3	23	605 mm. to 623 mm.
1887.												
January	7·5	-16·5	2	5	35	34	53	58	7	5	19	597 mm. to 626 mm.
February	11	-14	2	1	46	42	27	50	8	7	24	610 mm. to 628 mm.
March	21	8·6	2	1	55	45	35	70	16	7	34	599 mm. to 620 mm.
	16	-16·5	3	1	55	45	422	64	11	6	26	
1887.												
November	24·5	2	6	1	46	35	157	72	13	6	28	601 mm. to 621 mm.
December	14	-11	4	1	41	43	100	70	10	5	23	600 mm. to 617 mm.
1888.												
January	15	-21·5	2	1·5	45	45	30	60	10	7	26	601 mm. to 625 mm.
February	22	-14	3	2	55	45	320	80	16	3	30	597 mm. to 615 mm.
March	25	-9	4	2	62	47	300	80	19	5	35	598 mm. to 612 mm.
	20	-21·5	4	1.5	62	47	907	72	13	5	28	

SUMMARY FOR FIVE YEARS.

Date.	Temperature at 9 a.m.	Minimum temperature.	Cloud, 0 to 10.	Wind force.	Solar thermometer.	Maximum temperature.	Snow.	Percentage of relative humidity.	Weight of moisture in 10 cubic feet of air.	Drying power of the air.	Temperature at noon.
					Centgr.		ctm.				
1883-4	20·6	−7·5	4	0·9	61	45	146	69	12	5	26
1884-5	20·7	−10	4	0·9	61	53	269	70	13	5	28
1885-6	18·7	−14	4·3	0·9	50	49	245	69	12	5	26
1886-7	16	−16·5	3	1	55	45	422	64	11	6	26
1887-8	20	−21·5	4	1·5	62	47	907	72	13	5	28

Mean temperature of the day (9 a.m. to sundown) 27° Fahrenheit.

　　,,　　　　,,　　at 9.m.　　...　　...　　... 19°　,,

Average range from 9 a.m. to sundown　...　　... 29°　,,

Highest maximum　...　　...　　...　　...　　... 53°　,,

Lowest minimum　.　　...　　...　　...　　... − 21°　,,

Highest solar thermometer　...　　...　　...　　... 143°　,,

Weight of moisture in 10 cubic feet of air ...　　... 12 grains.

Drying power of air　...　　...　　...　　...　　... 5　,,

Drying power of the air at the temperature of the
　　human body　　...　　...　　...　　...　　.. 178　,,

HYGROMETRIC COMPARISON OF THE MALOJA WITH EGYPT.

	Mean Temperature.	Weight of Moisture in 10 cubic feet of Air.	Drying Power of Air at the Mean Temperature of the Atmosphere for 10 cubic feet.	Drying Power of Air at the Body Temperature, for 10 cubic feet.
	Fahr.	Grains.	Grains.	Grains.
* EGYPT.—				
January, 1884	50·5	30	11	160
February, ,, ...	54·5	33·6	14·4	156·4
November, ,, ...	63·5	40·3	23·7	149·7
December, ,, ...	58	38·3	15·7	151·7
January, 1885	54·5	32·6	15·4	157·4
February, ,, ...	57	33·2	18·8	156·8
Means ...	56·3	34·6	16·5	155·3
† MALOJA.—				
January, 1884	25·7	11	5·4	179
February, ,,	26·5	12·2	4·7	177·8
November, ,, ...	32·3	16·2	9·8	173·8
December, ,, ...	27	12·3	5	177·7
January, 1885	19·8	9·1	2·6	180·9
February, ,, ...	31·6	14·1	6·6	175·9
Means ...	27·1	12·5	5·6	177·5

* Calculated from eight observations during the twenty-four hours. (L'Observatoire Khédivial du Caire.)

† Calculated from the noon observations.

K

ATTRACTIONS

IN THE

NEIGHBOURHOOD OF THE MALOJA.

K 2

SUMMER EXCURSIONS

NEIGHBOURHOOD OF THE MALOJA.

No part of Switzerland offers a greater variety in attractions, easy excursions, and difficult mountain ascents. The elevation of the Engadine permits the ascension of some of the highest peaks and glaciers in Switzerland, and the region to the northward and westward of the Maloja Château is a part but little trod and almost unknown to tourists. The combination of excursions lying concealed behind these rugged slopes are incomparable in wildness and variety of scene, whilst the precincts of the Muretto Pass and Forno Glacier offer an endless field for exploration.

Until recently the banks of the Forno and Val Muretto were scaled merely by a few stragglers from the beaten tracks of Pontresina and St. Moritz. Even now the ponderous magnificence in the neighbourhood of Bacone, Del Forno, and Disgrazia, still lies hushed from common fame. The more demure loveliness of Bosco della Palza with innumerable wanderings lurking amongst green alps, fir trees, and luxuriant undergrowth, creates in the height of summer, delightful imaginings. The coolness of the sighing pines, and the tiny mountain ash, the clank of the cattle-bells, or splash of rising trout in the still morning waters. At every turn are lustrous colourings of sky, and lake, and rock-begirt groves sprinkled with rhododendrons.

There is no month, taken all round, that will compare with June in the diversities of natural attractions. The emerald freshness of a short grass garnished with myriads of little flowers, is only seen to perfection at this time. The air, too, is pungent and bracing.

What has invariably been considered an unsurmountable drawback to the June visit, is the possibility of a slight fall of snow. The Briton, from his point of view of what a snowfall entails, has always regarded with some misgivings a summer trip inclined to be so keenly tempered with winter memories, notwithstanding that it is very well known to *habitués* of the Engadine that a mountain snow-fall in summer is rare, and of quite a different character to that at low altitudes. June visitors are nevertheless increasing in numbers each year, and now that some proprietors of hotels render their houses comfortable by warming them during the cool evenings of June, no doubt many visitors who wish rather to escape the season's bustle, and at the same time secure a large share of the beauties of Nature, will be emboldened to face the overdrawn snow-spectre.

The chief objects of interest which surround the Maloja are as follows :—

The Forno Glacier, 2 hours distant. Take the road westward of the Kursaal in the direction of the ornamental châlets, pass the Osteria Vecchia, keeping on the main road until the Maloja Kulm is reached. In front of Hotel Kulm is a rock, with an iron railing on it, from whence a magnificent view may be obtained of the Bergell Valley. This point of view is only excelled by that seen from the Château Belvedere, on the hill to the right and hidden by the Mont des Chevres, to be described further on. Following the main road again to it first turning, a bridle path is taken in the direction of Piz Rossi (the peak with a small glacier on the face of it beyond which, and to the right, is seen a small snow-capped angle of the Cima di Rosso). Descend in a south-westerly direction, cross a small bridge over the river Ordlegna and follow the road to the left. In one hour the **Lac de Cavloccio**, is reached. This lake is remarkable for its placid waters and its curious green shade. A somewhat rugged path leads in the direction of the Muretto pass, and on emerging from the gorge, the Forno Glacier is seen on the right hand, the best path to which, must not be taken too high on the mountain side.

Monte Sissone (11,031 feet), from whence is a fine view of **Monte della Disgrazia.**

  Cima di Rosso (11,044 feet).

  Monte Forno (10,546 feet).

  Monte Muretto (10,476 feet).

  Monte della Disgrazia (12,057 feet).

All difficult, especially Disgrazia; but within ordinary reach of mountaineers.

Piz Margna (10,354 feet), with a steep glacier, 4 or 5 hours. A good view obtained of the snow-clad peaks of Monte del Forno, Rosatsch, Fex, Bernina, Morteratsch, &c.

Piz Corvatsch (11,342 feet), fronting Silvaplana, 9 or 10 hours. From the summit is seen an imposing view of the Bernina group, the valley of the Engadine, the Roseg Glacier, and nearly all the higher peaks of Switzerland.

Piz Surlej (10,456 feet), 7 or 8 hours.

Piz Julier (11,106 feet), 7 or 8 hours.

The Forno Pass. Over the Forno Glacier to St. Martino in the Val di Masino, 5 hours, difficult.

The Muretto Pass (8,389 feet). Approached by taking the route on the left bank of the River Ordlegna up to near the base of Monte Rossi, which we leave on the right, 8 or 9 hours to Chiesa. The Cavloccio route may be selected, and the river Ordlegna crossed by a rough bridge near the châlets of Piancanino at the termination of the débris and moraines of the Forno Glacier. After leaving Piancanino the path lies a little high, and to the right, but soon crosses the torrent when it rises to the left. The point of the glacier between Rossi and Forno is now traversed, the notch, in the distance, of the Muretto Pass, being unmistakable.

From Chiesa (4,282 feet), to Sondrio, the capital of the Valtellina, and from thence to Colico, returning to Maloja by Chiavenna and the Bergell Valley. Poschiavo can also be reached by crossing the Canciano Pass from Chiesa, or Sils, by the Tremoggia Pass (9,900 feet) and the Fex Glacier, or Pontresina, over the Scerscen and Capütschin passes.

The Val Malenco is well worthy of a visit, and also the Val Tellina, famous for its wine, of which there is so large a consumption in Switzerland.

The Muretto Glacier. Near the summit of the Pass, 3 hours from Lac Cavloccio. Splendid view of Monte della Disgrazia, Forno, Sissone, Rosso, &c.

Monte Salicina (8,498 feet), situated S.S.W. of Maloja, 4 or 5 hours distant. View obtained from the summit, of the Val Bregaglia and the mountains of Lombardy, &c.

Piz Lissone, a prominent double peak to the W.S.W., 5 to 6 hours distant. The Piz Duan is seen to the right and behind these peaks, the Gletscherhorn still further to the right.

The Septimer Pass (7,582 feet). One of the oldest historical routes travelled by the Romans. Descend from Maloja Kulm into the Bergell Valley and bear to the right behind Lunghino, crossing the Piano Maloggino on its old paved military road, or take the direction of Val Pila, crossing the stone avalanches of Longhino (guide necessary). The easiest and best route is from Casiccia, where a bridle path leads to the Passo di Sett—the junction of the Forcellina and Septimer passes—the latter joins the Julier route at Stalla (5,878 feet).

Piz Lunghino (9,100 feet), W.S.W. of the Kursaal where rise the Danube, the Po, and the Rhine ; 3 hours from Maloja. The mountain is of peculiar interest to anyone who is equal to three hours climbing, for it constitutes a portion of the water-parting, from which it can be said that three large rivers take their sources. The origins of the Inn, the Maira, and the Oberhalbstein Rhine are found here. These small rivers respectively flow into the Danube, the Po, and the Rhine.

The higher peak is reached in about three hours from the hotel. An ascent can be made by the winding bridle path in Val Pila, constructed last summer up to the Lunghino Lake ; or the gentler decline taken by the path at the edge of the Maloja Lake near Capolago. The peak is attained by skirting the Lunghino Lake on its S.W. border, and ascending the N.W. slope. From this point at the edge of the Lake is seen the path of—

Fourcla da Lunghino. A pass leading into the Septimer and Forcellina to the Avers Valley. A return journey can also be made by descending the Septimer path to Casaccia and mounting the Maloja pass.

Piz Gravasalvas (9,620 feet), 4 or 5 hours, difficult. To the N.W. of the Kursaal. Septimer Pass to the W., and N.

Motta Rotondo (8,100 feet), 3 hours. To the N. of the Kursaal. The greenest grass reaches to its summit. The effect of verdant patches of short thick herbage, interspersed with the red tint of the Alpine rose, adds a peculiar fertile glow to these rough slopes.

Piz Materdell (9,700 feet), 4 or 5 hours distant. N.N.E. of the Kursaal.

Piz Lagrèv (10,398 feet). The rugged pointed peak to

K α

the N.E., very difficult and dangerous to mountaineers unac-
customed to rock-work. 7 or 8 hours.

Piz Pulaschin (9,898 feet), N.E. of Lagrev, 5 or 6 hours.

From many of the peaks can be seen a panorama of entire
Switzerland—Mont Blanc, Todi, Russein, Finsteraahorn,
Monte Rosa, Jungfrau, &c.

If desired, carriages conduct one near the paths of ascent,
and, with the assistance of guides, the interesting points of
view can be attained. For delicate persons or others who do
not care to make ascents, more or less difficult, there are in the
neighbourhood, several pleasant walks, viz. :—

Casaccia. A small village at the foot of the Septimer
Pass. Descend the zig-zag road after passing Maloja Kulm.
The ruins of a church, St. Gaudenzio, said to have been constructed in the fourteenth century, are seen on the right,
before entering the village. Time required from Maloja to
Casaccia about 50 minutes. A short cut to Casaccia is made
by the gap to the right of the rocky eminence W. of the
Maloja Kulm. A romantic winding track takes the
pedestrian into the main road low down.

The Cascade de l'Ordlegna, 25 minutes distant. When
illuminated at night produces a very pretty effect. Situated
half-way down the zig-zag road to Val Bregaglia, a directing
post indicates the path to the Falls, which are 5 minutes
distant from the main road.

Chateau Belvedere, 15 minutes distant. Take the road
to the N.W. of the Osteria Vecchia and follow its windings.
On the left the Val Murretto and Muretto Pass, leading to
the Val Malenco (8 or 9 hours) and Val Tellina. In front
the picturesque Bergell Valley enclosed by high jagged
mountains. To the right the yawning Val Marozzo and
Piz Lunghino. This solid structure, built in the mediæval
style of a castle, stands higher than the kulm of St. Moritz.
The ground is at an elevation of 213 feet above the Maloja
Lake, and commands one of the most charming views in the

whole of Switzerland, far surpassing anything to be seen in the Engadine. The Château stands on an isolated hill (la Petite Colline) covered with the *pinus montana*. The wild rocks, rhododendrons, grasses, and fir trees on the south-west face of this huge balcony, give it a fertile, pleasing effect, whilst the surroundings are massive and grand. From the "Knoll"—a pine-covered rock to the west of the Chateau—is an enchanting panorama of grandeur, in the view down on the bed of the river Ordlegna, sweeping the tips of the firs of Piano Maloggina and the woods of Canova and Cavrille, in the dip of over 1,400 feet to Casaccia, the highest village in the Bergell Valley. It is down the winding road on the left that the winter runs are made on the Canadian sled over the polished snow, in less than eight minutes to the plain below.

Chemin du Artistes. This delightful walk is on the western side of the Belvedere, winding between huge rocks and pines away to the Val Pila (45 minutes). The most varying and romantic scene is passed, and many objects of interest present themselves, amongst which are the Gletscher-muhle, the track of the stone-avalanches from the Lunghino, the Falls of the Inn, the serrated peaks of Val Bregaglia, and the zig-zag road leading to Casaccia.

Colline du Château. This hill is approached by crossing the wooden bridge to the right of the Belvedere. On its summit a different view is obtained of all the points of interest mentioned in the last route, as well as a view of the Maloja Lake, stretching away to Sils Maria and shut in by the Bernina and Albula chains.

Val d'Enfer, approached from the main road, crossing the grass on the right, after passing the dogano.

Lac de Cavloccio, 50 minutes *ride* route to Forno Glacier

Lac Bitabergo (6,107 feet), situated below the rocky brow of Salecina south of the Kursaal. The Cavloccio road is taken as far as an irregular rough plateau covered with pines, underwood, and rocks; to the right of this a hollow depression on the mountain side leads to the lake.

Sils Baselgia, 50 minutes. A little village, situated in a wild and picturesque spot, at the N.E. end of the lake.

Sils Maria, 1 hour. At the entry of the valley of Fex, surrounded by hills covered with larches and red pines.

VIEW FROM SILS MARIA ON MALOJA.

Crestalta, about 2 hours distant. Extensive view from the summit of a cone of the lakes of Sils and Silvaplana, with the Maloja plateau on one hand and St. Moritz on the other.

Chasté. In front of Maloja at the opposite end of the lake, on an islet where are found the ruins of an ancient castle.

Isola. A small hamlet seen on a green plateau facing

Maloja ; on the S.E. side of the lake. The falls formed by the water from the Fedoz Glacier are five minutes from the village. A quaint old house, with curious paintings on the walls, is open to strangers for view. Take the road S.E. of Kursaal, passing the group of châlets named Cresta. The walk, a very beautiful one, skirting the lake, amongst pine-trees, boulders, and wild shrubberies, will occupy about an hour.

Crap da Chüern. An enormous perpendicular rock, crowned with forest, dominating the lake, and forming a high precipice. It may be ascended from the road by passing to the left and behind. Time required, about $1\frac{1}{4}$ hours from the hotel.

Platz de la Pininsule, 25 minutes distant. Situated at the base of the Crap da Chüern, covered with rocks and pines, Fedoz Glacier to the south. Following the main road, the little valley of **Gravasalvas** opens out from the Lake of Sils. The villages of Gravasalvas and Blaunca are situated here, and Lac Nair (8,000 feet) is gained from the valley.

Val Pila, 15 minutes. A hidden green valley at the base of Longhino, the path to which is on the right hand side of the quarry. A return journey can be made by traversing the valley and mounting towards the Belvedere.

Lac Lunghino (8,124 feet) 2 hours. Situated to the N.W. of the Hotel Kursaal, in the saddle formed by Piz Lunghino and Piz Gravasalvas.

Lac Nair (8,000 feet). May be reached from the main road in the direction of the Crap da Chuern. A path will be discovered on the left, about half a mile from the hotel (*vide* Gravasalvas).

Fedoz Glacier. Between Piz Margna and Piz Güz (10,397 feet) is seen from near Isola and from the main road leading to Sils.

Fex Glacier. In continuation with the Val Fex at Sils-Maria. A carriage takes one near the Glacier.

THE VALLEY OF FEX.

Ordeno, 30 minutes. A group of chalets situated on a grassy plateau overlooking the banks of the river Ordlegna. The continuation of the route on this side of the river leads to the Muretto Pass.

Albigna Water-fall. Seen from below Casaccia before entering Vicosoprano.

Gravasalvas, 1 hour. After passing Capolago, the village in front of the Kursaal, a road branches off at the first rivulet which leads to Gravasalvas. This route is easily found, and consists in its first part of an old Roman road, cut in many places out of the solid rock overlooking the lake. On the road are Spluga and Buera, hamlets secreted from the outer world in the massive clefts of the Albula range. By branching off at Buera, the second village, and pursuing the track along the banks of the Ova del Mulin, Blaunca is reached, and following the river still further, Lac Nair. From the small cluster of habitations named Blaunca, there is also a mountain path to Gravasalvas and from thence to the main road, where the return to the Maloja will occupy about three-quarters of an hour. The village and valley of Gravasalvas must not be confounded with the Piz Gravasalvas which is situated to the westward 3,000 feet above the valley of the same name.

L'Ala. A small mount 1,196 feet above the level of the lake situated to the S.E. of the Hotel. The summit can be reached in one hour by taking the path to the left of the church and gaining the eastern slope of l'Ala. From its rocky eminence a full view is obtained of the Forno Glacier, the Cavloccio Lake, Val Bregaglia, and lakes of the Engadine. Facing l'Ala in a direct line with the Hotel is the Motta Rotondo.

A short promenade of 30 minutes can be made by taking the path on the S.E. of the Kursaal, passing through the hamlet of Cresta, and making for the Church, after a few windings one emerges by the Osteria Vecchia. Longer walks in the direction of St. Moritz, or down the Val Bregaglia towards Promontogno, are available, and in winter the descent on a toboggan to Vicosprano, and even Promontogno (14 miles) is easily made. The charms of these winter excursions in the glistening snow and brilliant sunshine, with deep blue sky, are unknown to summer visitors. In winter, also skating can be had, both on the Maloja Lake (3 miles long), and on the ice rink at the rear of the Hotel.

Bosco della Palza. The wood within a few minutes walk, to the E. of the Kursaal. The main path if followed for an hour leads to Isola.

Val Bregaglia. Commences at the foot of the Maloja

Pass, and extends for about 18½ miles to Chiavenna. The
Italian frontier is passed at Castasegna (13 miles from
Maloja).

On either side of the valley are towering crags and
needle-shaped peaks, the S.W. side being most imposing in
its rude stateliness. It is said that the spade-shaped mountain
—Piz Badil—has been ascended but once. In 1884, two
Englishmen with guides had to turn back from the difficulties
of the ascent.

The contrast of the Val Bregaglia with those more expanded
and massive beauties of the Engadine, enhances the effect
caused by the abrupt change of scene, when arriving at the
summit of the Maloja Kulm.

Distances.

Chiavenna to Samaden by the Val Bregaglia and Maloja
Pass :—

					Kilomètres.	Miles.
Chiavenna	-	-	-	-	0	0
Villa	-	-	-	-	7·2	4½
Castasegna	-	-	-	-	9·6	6
Promontogno	-	-	-	-	13·2	8
Stampa	-	-	-	-	16·3	10
Borgonouvo	-	-	-	-	17·5	10½
Vicosoprano	-	-	-	-	19·0	11½
Casaccia	-	-	-	-	26·2	15¾
MALOJA (Kulm)	-	-	-	31·1	18½
Sils	-	-	-	-	38·0	4½
Silvaplana	-	-	-	-	42·3	7½
Campfèr	-	-	-	-	44·7	9
St. Moritz	-	-	-	-	48·9	11¼
Celerina	-	-	-	-	52·0	13
Samaden	-	-	-	-	54·4	14¼

A Kilomètre equals 0·6213924 miles, or 1093.6331 yards, roughly 3-5ths
of a mile.

INDEX.

www.ingramcontent.com/pod-product-compliance
Lightning Source LLC
Chambersburg PA
CBHW021804190326
41518CB00007B/450